LOS ANGELES COUNTY, CHAPTER

The National Electrical Contractors Association is an organization dedicated to improving the quality of electrical service available to American industry and the general public. For more than three quarters of a century NECA has developed and compiled a tremendous amount of data and information on the many facets of the electrical construction industry.

This material, derived from the experience of the thousands of electrical contractors, is made available through the Association in a wide variety of publications and programs. Electrical contractors drawing from this vast storehouse of knowledge and experience may use the information to enhance their own abilities, to advance the industry and, in the final analysis, to benefit the consuming public.

In addition, through NECA, the interests of the electrical contractor are represented to all branches and levels of the government, to other segments of the construction and electrical industries and to the public generally. In these many ways NECA is the focal point for the electrical contracting industry in its continuing efforts to improve the quality of the nation's electrical service.

TABLE OF CONTENTS

N.E.C. CHAPTER 1. GENERAL

N.E.C. CHAPTER 2. WIRING AND PROTECTION

N.E.C. CHAPTER 3. WIRING METHODS AND MATERIALS

N.E.C. CHAPTER 4. EQUIPMENT FOR GENERAL USE

400 ***FLEXIBLE CORDS AND CABLES***

N.E.C. CHAPTER 5. SPECIAL OCCUPANCIES

N.E.C. CHAPTER 6. SPECIAL EQUIPMENT

N.E.C. CHAPTER 7. SPECIAL CONDITIONS

APPENDIX

ELECTRICAL SERVICE ENTRANCE AND BONDING DIAGRAMS

LOAD CURRENT CALCULATIONS

LOCKING TYPE PLUGS

N.E.C. Article 100 — Definitions

Plenums

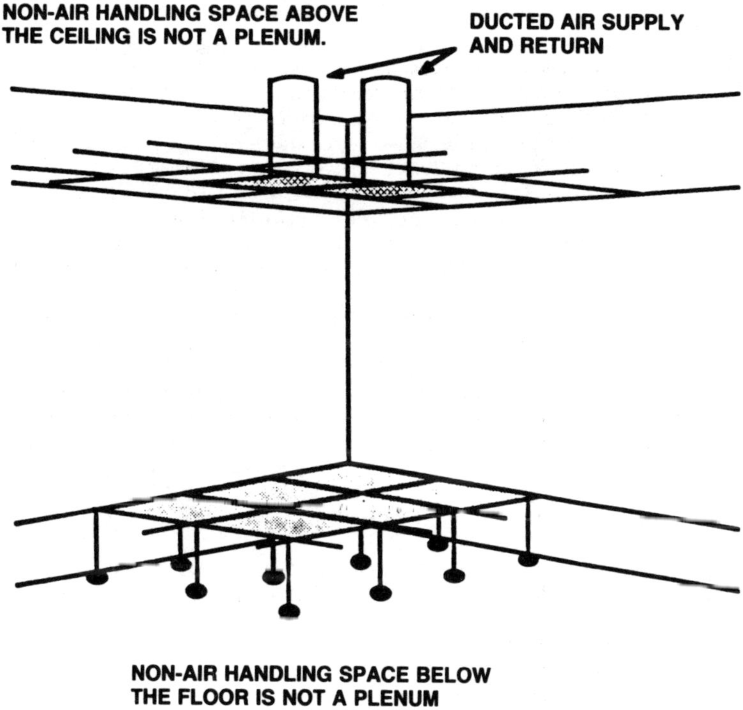

N.E.C. Section 110-3 (b)

Rules For Torquing

SECTION 110-3 (b) REQUIRES (A MANDATORY RULE) ALL LISTED AND LABELED PRODUCTS TO BE USED IN ACCORDANCE WITH ANY INSTRUCTIONS THAT ARE GIVEN IN A TESTING LABORATORY'S LISTING BOOK, SUCH AS UL GREEN BOOK OR WHITE BOOK. AFTER OCTOBER 1, 1986 ALL FIELD TERMINATIONS FOR UL STANDARD LISTED BELOW, SHALL BE TORQUED (NOT OVERTORQUED) TO THE VALUES RECOMMENDED BY UL OR THE MANUFACTURERS:

UL — 486	CONNECTORS	UL — 845	MOTOR CONTROL CENTERS
— 67	PANELBOARDS	— 857	BUSWAYS
— 98	ENCLOSED SWITCHES	— 869	SERVICE EQUIPMENT
— 231	POWER OUTLETS	— 891	SWITCHBOARDS
— 414	METER SOCKETS	— 1008	AUTO. TRANSFER SWITCHES
— 489	MOLDED CIRCUIT BREAKERS	— 1047	ISOLATION SWITCHES
— 509	INDUSTRIAL CONTROL EQUIP.	— 1059	TERMINAL BLOCKS
— 512	FUSEHOLDERS	— 1062	UNIT SUBSTATIONS

IF THE CONNECTOR IS A MECHANICAL SCREW TYPE, APPLY THE MANUFACTURER'S RECOMMENDED TORQUE. IN ABSENCE OF SPECIFIC TORQUE RECOMMENDATIONS USE UL486B TORQUE VALUES.

APPLY PROPER TORQUE TO SCREW TYPE CONNECTORS

TORQUE FOR SOCKET HEAD SCREWS[a]

Socket Size Across Flats – Inches	Torque Pound – Inches
1/8	45
5/32	100
3/16	120
7/32	150
1/4	200
5/16	275
3/8	375
1/2	500
9/16	600

TIGHTENING TORQUE FOR SCREWS[a]

	TORQUE, POUND-INCHES			
	Slatted Head No. 8 and Larger[b] Slot Width - Inches		Hexagonal Head External Drive Socket Wrench	
	To 3/64	Over 3/64		
Wire Size	Slot Length - Inches To 1/4	Over 1/4	Split-Bolt Connectors	Other Connectors
8 AWG	25	40	80	75
6	35	45	165	110
4	—	45	165	110
3	—	50	275	150
2	—	50	275	150
1	—	50	275	150
1/0	—	50	385	180
2/0	—	50	385	180
3/0	—	50	500	250
4/0	—	50	500	250
250 MCM	—	50	650	325
300	—	50	650	325
350	—	50	650	325
400	—	50	825	325
500	—	50	825	375
600	—	50	1000	375
700	—	50	1000	375
750	—	50	1000	375
800	—	50	1100	500
900	—	50	1100	500
1000	—	50	1100	500

a. CLAMPING SCREWS WITH MULTIPLE TIGHTENING MEANS: FOR EXAMPLE, FOR A SLOTTED HEXAGONAL HEAD SCREW, USE THE HIGHEST TORQUE VALUE ASSOCIATED WITH THE DIFFERENT TIGHTENING MEANS.

b. FOR VALUES OF SLOT WIDTH OR LENGTH OTHER THAN THOSE SPECIFIED, SELECT THE LARGEST TORQUE VALUE ASSOCIATED WITH CONDUCTOR SIZE.

NOTE: THE TORQUE TABLES PRESENTED HERE ARE TAKEN FROM UL STANDARD 486B, BUT ARE REPRESENTATIVE OF THOSE PUBLISHED IN OTHER UL STANDARDS, IN NEMA EQUIPMENT INSTALLATION INSTRUCTION PUBLICATIONS, AND IN THE CANADIAN ELECTRICAL CODE. THE SAME VALUES APPLY TO PRESSURE CONNECTORS FOR BOTH COPPER (UL STANDARD 486A) AND ALUMINUM CONDUCTORS.

N.E.C. Section 110-3 (b)

Listing Of Equipment

EVERY MATERIAL, FITTING, DEVICE, APPLIANCE, APPARATUS OR EQUIPMENT FOR GENERAL USE SHALL BE LISTED* AND APPROVED** BY A RECOGNIZED TESTING LABORATORY.

CONSULT YOUR LOCAL AGENCIES FOR APPROVED LABORATORIES

ALL EQUIPMENT SHALL BE LISTED FOR THE SPECIFIC PURPOSE FOR WHICH IT IS USED

FACTORY MUTUAL (FOR HAZARDOUS LOCATION EQUIPMENT ONLY).

*In City of Los Angeles listing is mandatory (LAMC 93.0402) and
**Testing agencies must be approved by Dept. of Building and Safety (LAMC 93.0503).

N.E.C. Section 110-14 (a)

Limitations Of Conductors On Receptacles Terminals

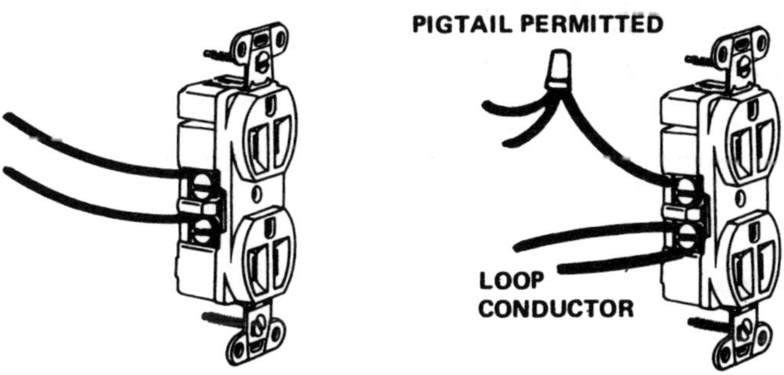

O.K. EACH SCREW IS CONSIDERED A SEPARATE TERMINAL.

NOTE: A LOOP CONDUCTOR IS CONSIDERED 1 CONDUCTOR IN MULTIWIRE CIRCUITS THE CONTINUITY OF AN IDENTIFIED GROUNDED CONDUCTOR SHALL NOT BE DEPENDENT UPON DEVICE CONNECTIONS.

N.E.C. Section 110-14 (a)

Terminating Stranded Wire

NOT APPROVED

STRANDED CONDUCTORS TERMINATED WITHOUT SOLDER OR PRESSURE LUG

N.E.C. Section 110-14 (b)

Pressure Cable Connector Used With A Combination Of Conductors Not Specified In Approval

ALL SPLICING DEVICES MUST BE IDENTIFIED FOR THE USE, INCLUDING CONDUCTOR MATERIAL

NOT APPROVED

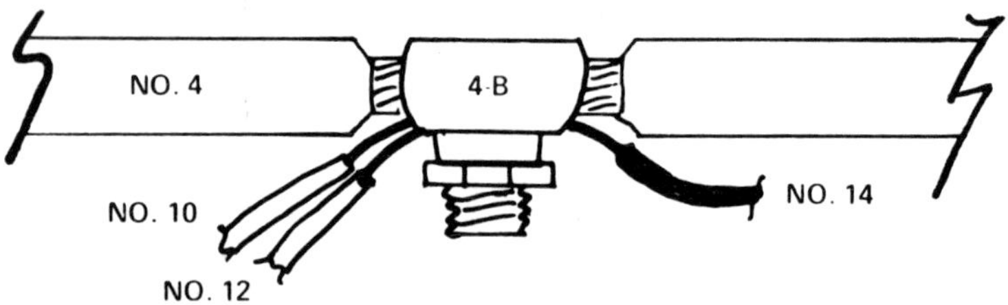

DATA ON CARTON

PRESSURE CABLE CONNECTORS				
CAT. NO.	TWO WIRES SAME SIZE AWG - MCM	RUN AND TAP SIZES		
		RUN	MAX TAP	MIN
4B	4	4	4	8
6B	8 TO 6	6	6	10
1D	6 TO 2	2	3	14
4D	8 TO 4	4	6	14

U.L. NOTE: PRESSURE CABLE CONNECTORS ACCOMMODATE TWO SOLID AND/OR STRANDED WIRES UNLESS OTHERWISE SPECIFIED.

N.E.C. Section 110-16 (a)

600 Volts Or Less

N.E.C. Section 110-34 (a)

Over 600 Volts

THE MINIMUM CLEAR LEVEL, UNIMPAIRED STANDING AND WORKING SPACE DIRECTLY IN FRONT OF ELECTRICAL EQUIPMENT WHICH REQUIRE EXAMINATION, ADJUSTMENT, TESTING OR REPAIRING OR WHERE LIVE PARTS ARE EXPOSED SHALL NOT BE LESS THAN LISTED AS FOLLOWS:

VOLTAGE TO GROUND	LIVE PARTS MEASURED AT RIGHT ANGLES TO THE FACE, COVER OR DOOR OF THE ENCLOSURE UNGROUNDED OPPOSING SURFACE	LIVE PARTS (WHERE PERMITTED) MEASURED AT RIGHT ANGLES FROM THE FURTHEST PROJECTION OF THE EXPOSED LIVE PARTS GROUNDED OPPOSING SURFACE	EXPOSED LIVE PARTS BOTH SIDES WITH OPERATOR IN BETWEEN
0– 150	–3 FT –	– 3 FT.	– 3 FT.
151– 600	–3 FT. –	– 3½ FT.	– 4 FT.
601–2500	–3 FT. –	– 4 FT.	– 5 FT.
2501–9000	–4 FT. –	– 5 FT.	– 6 FT.
9001–25000	–5 FT. –	– 6 FT.	– 9 FT.
25001–75KV	–6 FT. –	– 8 FT.	– 10 FT.
ABOVE 75KV	–8 FT. –	–10 FT.	– 12 FT.

N.E.C. Section 110-16 (a)

Working Space Required For Motor Controller

NOT APPROVED

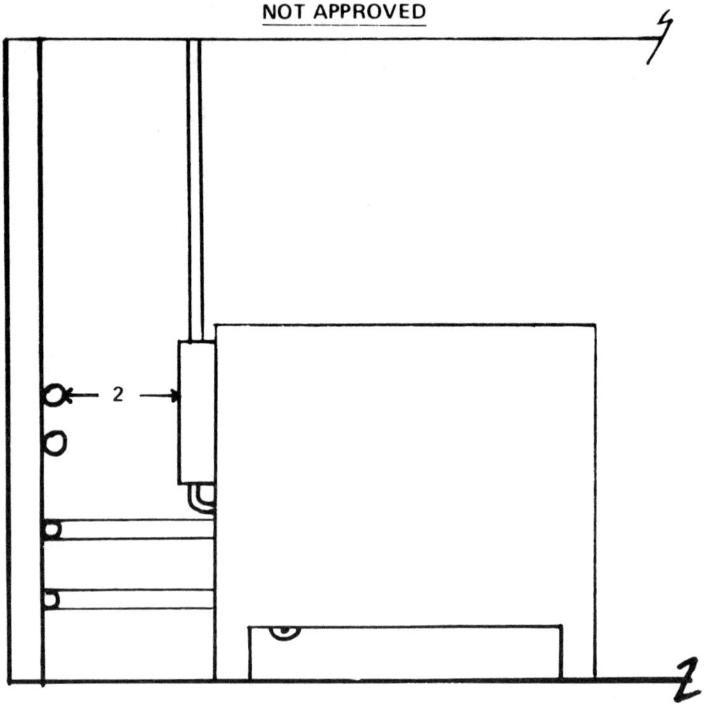

MOTOR CONTROLLER WITHOUT
MINIMUM 3′ WORK SPACE.

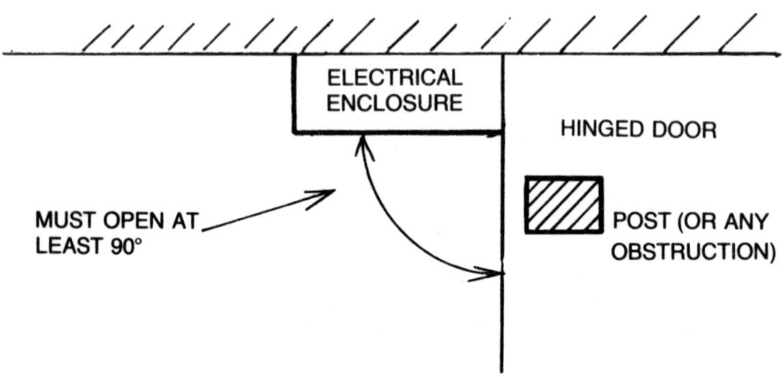

N.E.C. Section 110-16 (c)

Minimum Access And Entrance To Working Space

EQUIPMENT RATED 1200 AMPS AND OVER 6 FT. WIDE REQUIRED TO HAVE A 24" ENTRANCE AT EACH END

APPROVED

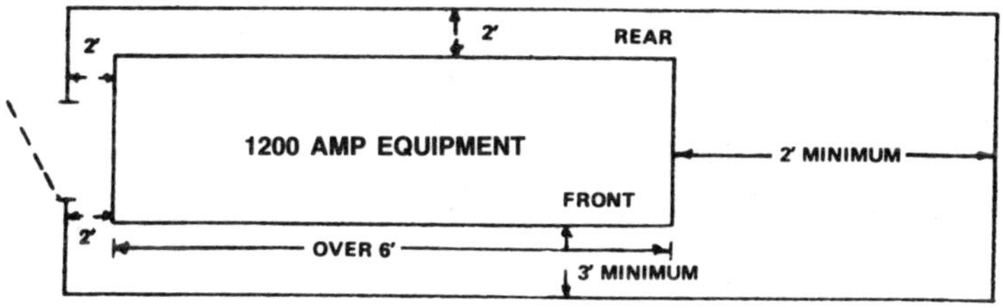

TABLE 110-16(a) OR 110-34(a) FOR MINIMUM DISTANCE

NOT APPROVED

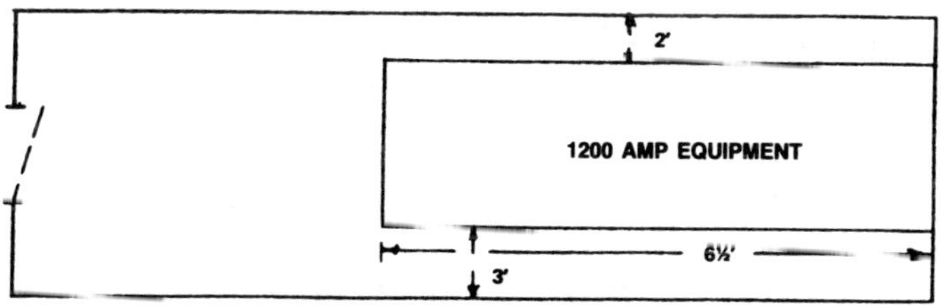

NO ACCESS AT END OF EQUIPMENT

NEW EXCEPTIONS 1 & 2

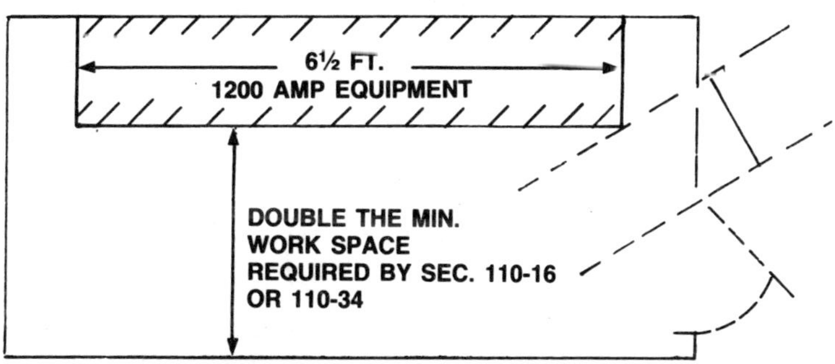

DIMENSION X MUST BE EQUAL TO DISTANCE GIVEN IN TABLE 110-16(a) OR 110-34(a)

N.E.C. Section 110-16 (f)

Headroom Required For Switches And Controllers

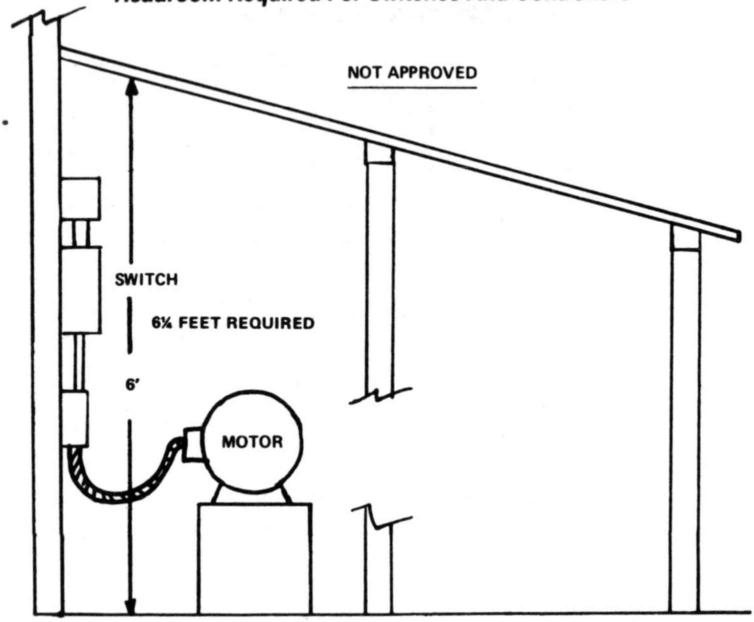

MOTOR CONTROLLERS AND EXO SWITCHES NOT APPROVED WITH LESS THAN 6¼' HEADROOM.

N.E.C. Section 110-17

Guarding Of Live Parts

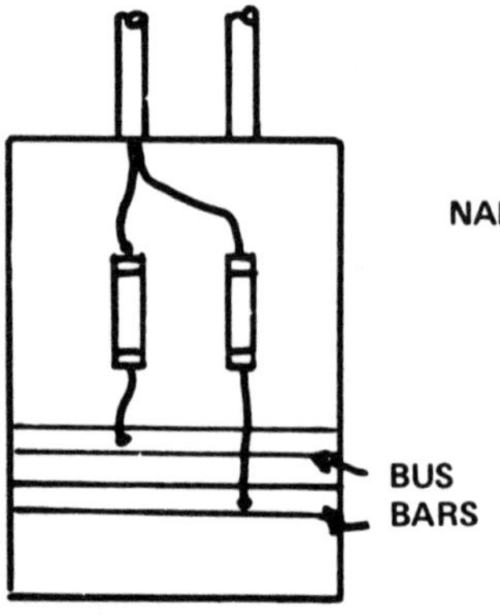

NOT APPROVED

"LIVE-BACK" MOTOR CONTROLLER.
NEW INSTALLATION

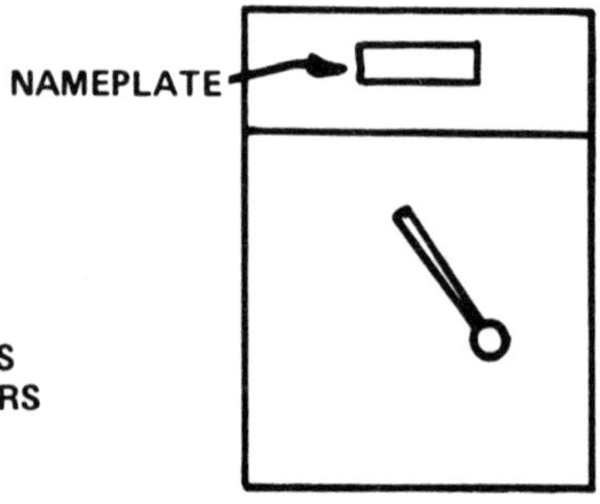

APPROVED

TOTALLY ENCLOSED CONTROLLER, SWITCH, OR CIRCUIT BREAKER. NAME PLATE TO BE VISIBLE AT ALL TIMES.

N.E.C. Section 110-22

Identifying Equipment

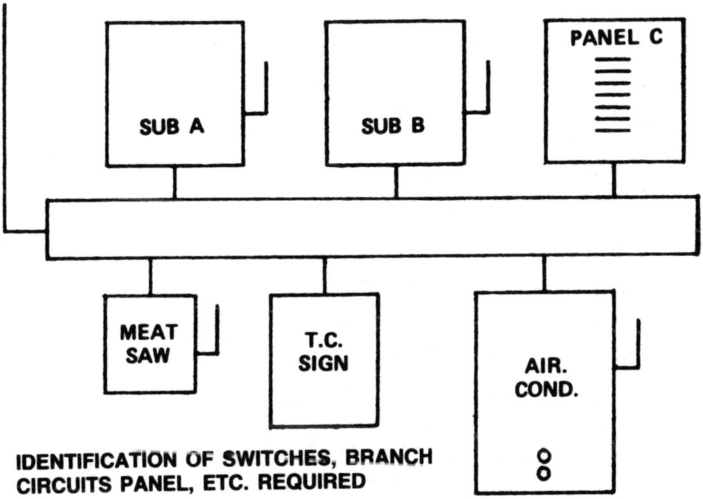

N.E.C. Section 110-31

Equipment Over 600 Volts

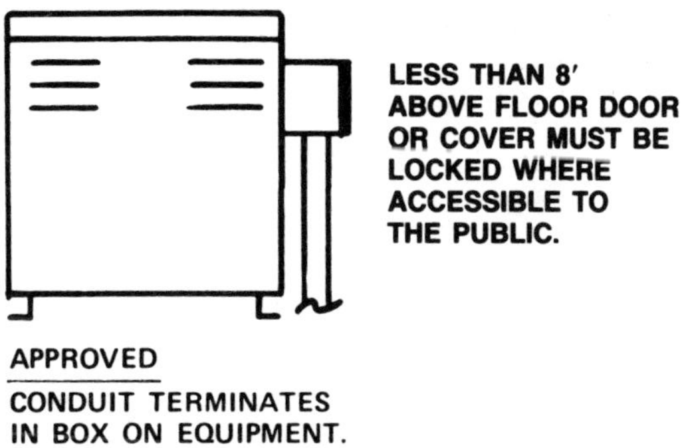

N.E.C. Section 110-34 (e)

Unguarded Live Parts

(OVER 600 VOLTS)

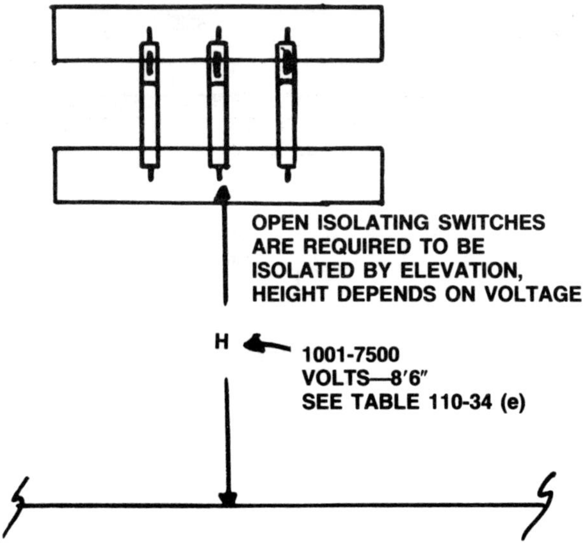

N.E.C. Section 200-6(d), 210-5

Grounded Conductors of Different Systems

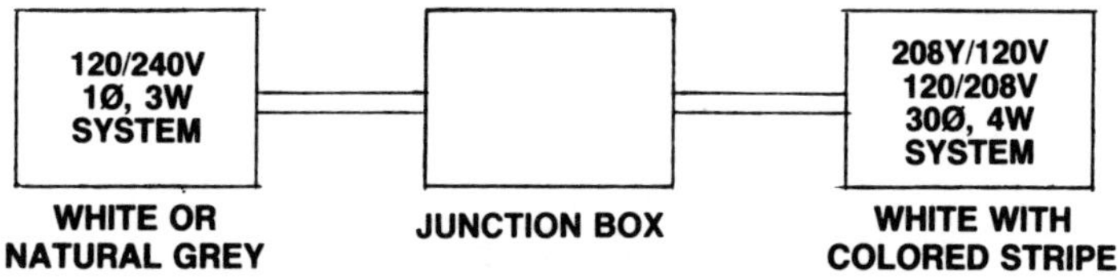

GROUNDED CONDUCTORS OF DIFFERENT SYSTEMS MUST BE IDENTIFIED DIFFERENTLY.

N.E.C. Section 210-1

Article 430 Applies To Motor Circuits

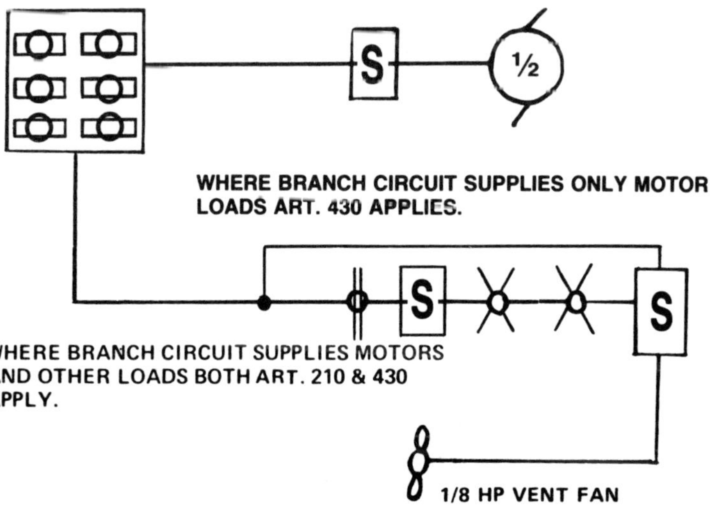

N.E.C. Section 210-4

Multi-Wire Branch Circuits

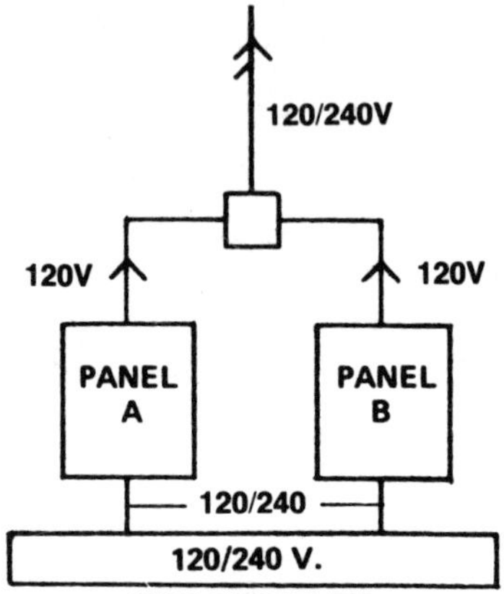

NOT APPROVED

CONDUCTORS OF A SINGLE MULTI-WIRE BRANCH CIRCUIT SUPPLIED FROM TWO PANELBOARDS

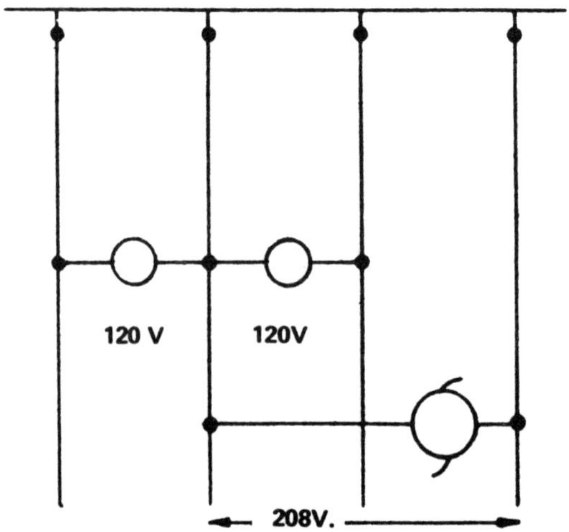

UNEQUAL VOLTAGE BETWEEN UNGROUNDED CONDUCTORS AND GROUND WIRE IS NOT DEFINED AS A MULTI-WIRE BRANCH CIRCUIT (SEE DEFINITION ARTICLE 100)

N.E.C. Section 210-4 (c) (Ex. 1 & 2)

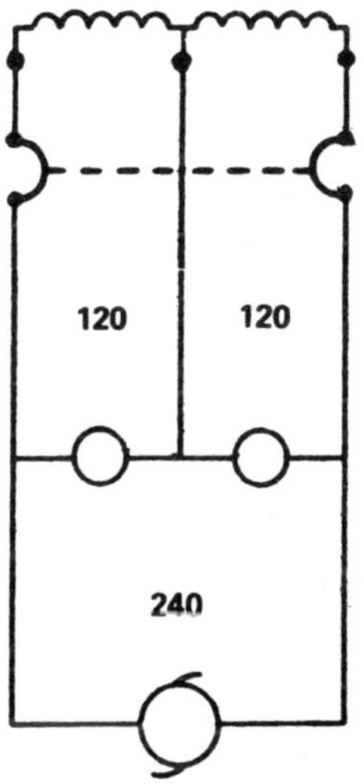

APPROVED
MULTI-WIRE CIRCUITS
SUPPLYING 120V
AND 240V LOADS WHERE
PROTECTED BY A SINGLE
2 POLE CIRCUIT BREAKER
OR
SUPPLYING ONLY ONE
UTILIZATION EQUIPMENT

N.E.C. Section 210-4 (d)

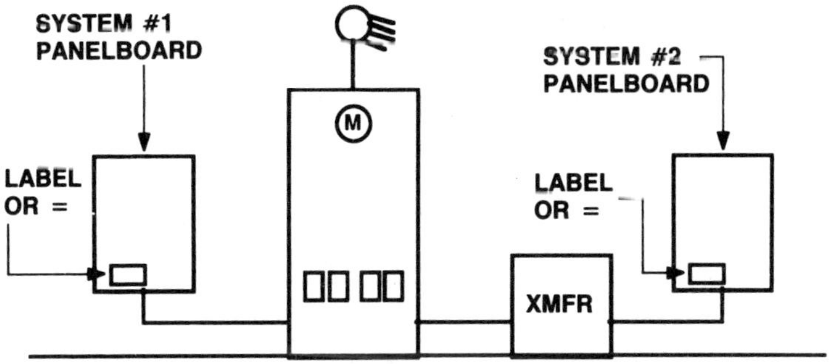

TWO NOMINAL VOLTAGE SYSTEMS IN ONE BUILDING MUST HAVE ALL UNGROUNDED CONDUCTORS IDENTIFIED BY PHASE AND SYSTEM. THE MEANS OF IDENTIFICATION SHALL BE POSTED AT EACH BRANCH CIRCUIT PANELBOARD

N.E.C. Sections 210-4 (c) (Ex. 2) & 240-20 (b) Ex.

Ungrounded Conductor Of Multiwire Branch Circuits

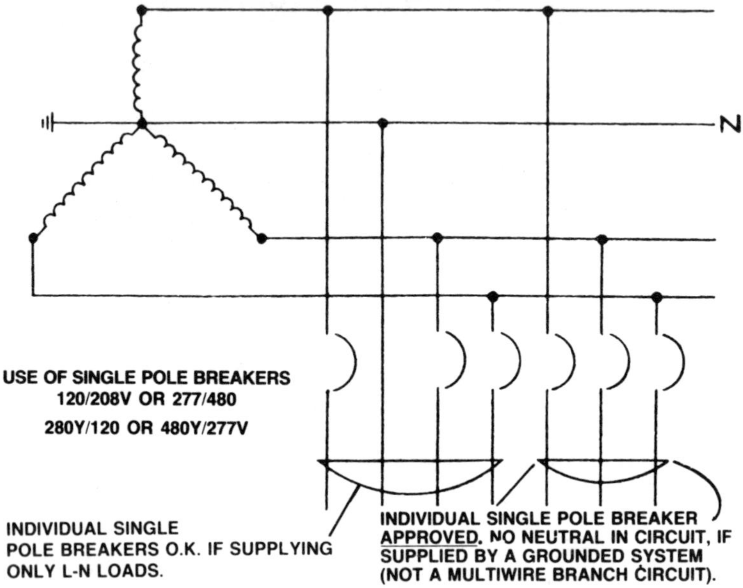

N.E.C. Section 210-4 (b)

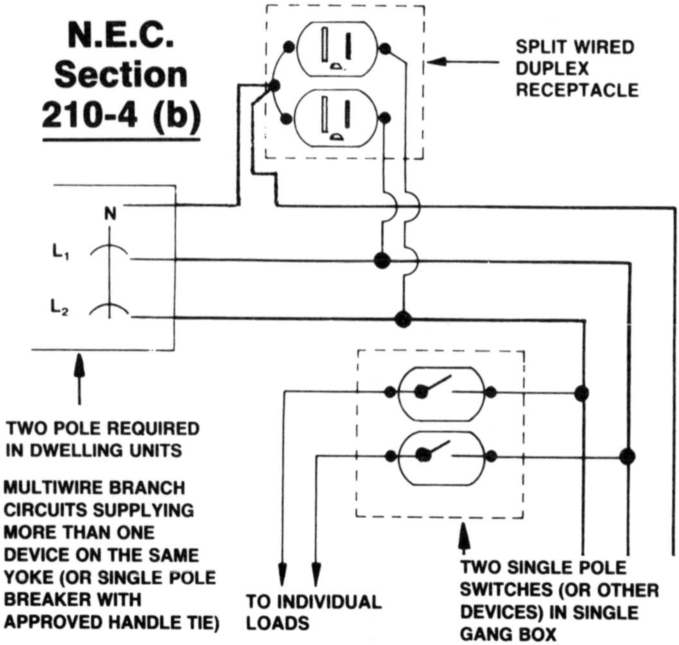

N.E.C. Section 210-6 (a) and (c)

Maximum Voltage For Lampholders

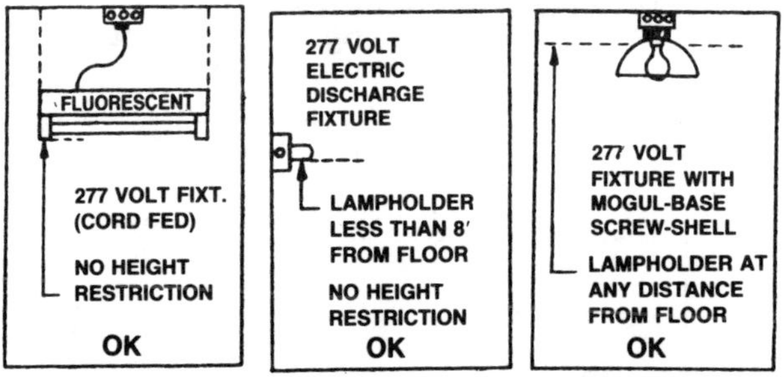

210-6 (c) 3 | 210-6 (c) 1 | 210-6 (c) 2

ABOVE DRAWINGS FOR INSTALLATION IN INDUSTRIAL BUILDINGS, OFFICE BUILDINGS, SCHOOLS, STORES, AND PUBLIC & COMMERCIAL AREAS OF OTHER BUILDINGS SUCH AS HOTELS OR TRANSPORTATION TERMINALS. THE HEIGHT TO A PERMANENTLY INSTALLED ELECTRIC DISCHARGE LAMP OF THE SCREW SHELL TYPE MAY BE LESS THAN 8′ FROM THE FLOOR.

N.E.C. Section 210-6 (d) (1) a

Voltage Between Conductors Over 277 Volts And Under 600 Volts

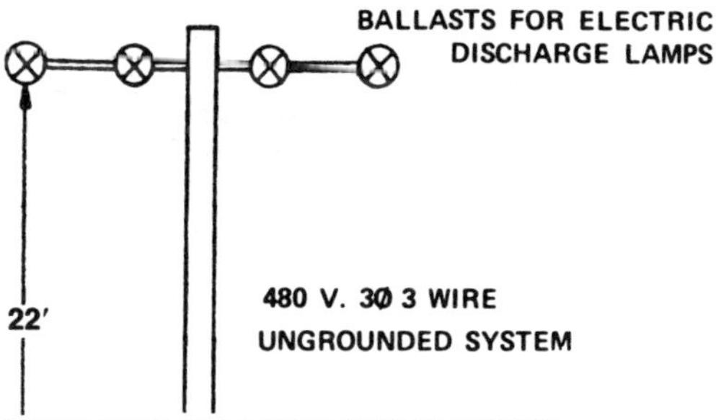

BRANCH CIRCUITS SUPPLYING BALLASTS FOR ELECTRIC DISCHARGE LAMPS PERMANENTLY INSTALLED FOR ILLUMINATION OF AREAS SUCH AS BRIDGES, ATHLETIC FIELDS, PARKING LOTS AT A HEIGHT NOT LESS THAN 22 FT. SHALL NOT EXCEED 600 VOLTS BETWEEN CONDUCTORS. SEE SECTION 410-78.

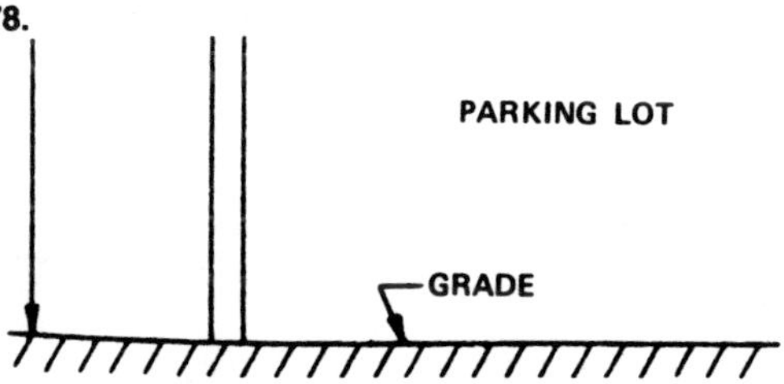

N.E.C. Section 210-7 (a)

Required Type Of 120 Volt General Use Receptacle

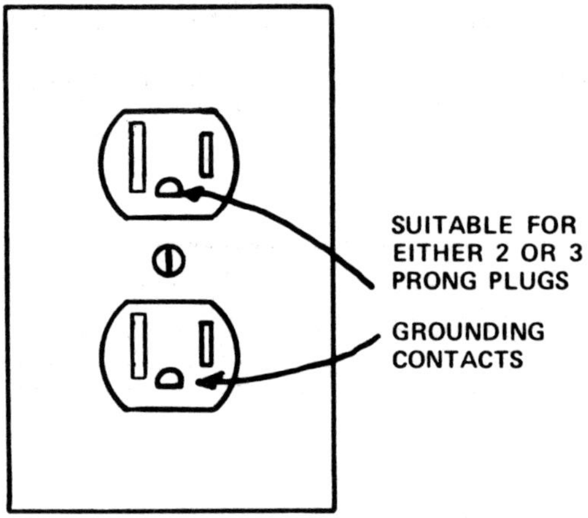

N.E.C. Section 210-7 (d)

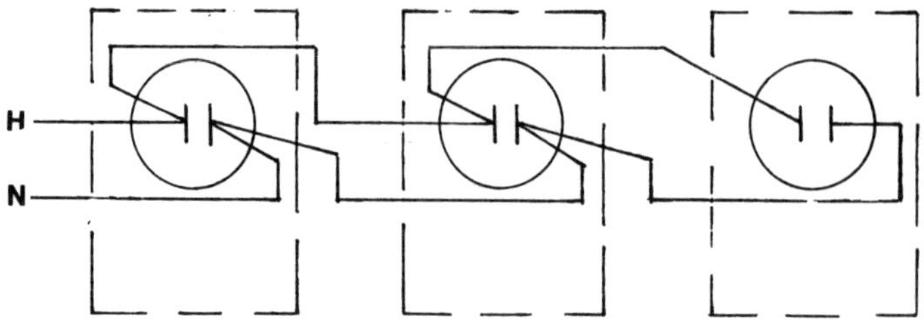

THIS EXISTING NON-GROUNDING DEVICE IN BOX WITHOUT AN EQUIPMENT GROUND MAY BE REPLACED BY A GFCI RECEPTACLE AND MAY FEED DOWNSTREAM RECEPTACLES, WHICH MAY THEN BE OF THE GROUNDED TYPE.

N.E.C. Section 210-8

Ground-Fault Circuit Protection

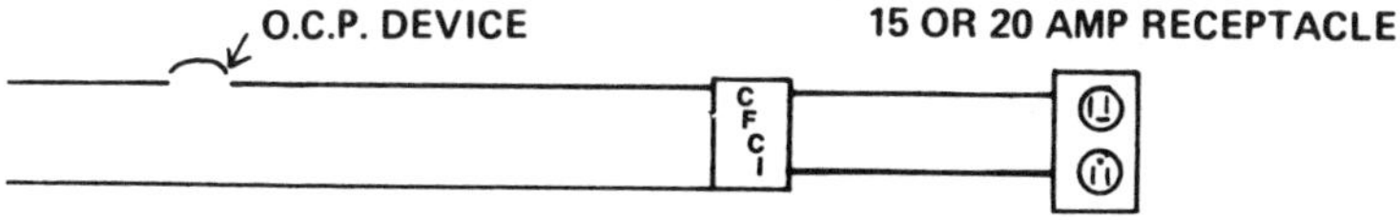

210-8 (a) DWELLING UNITS
210-8 (b) HOTELS AND MOTELS

1) **WITHIN 6 FT. RADIUS FOR RECEPTACLES TO SERVE COUNTERTOP SURFACES IN A DWELLING UNIT KITCHEN.**
2) **CONSTRUCTION SITES, SEE 305-6.**

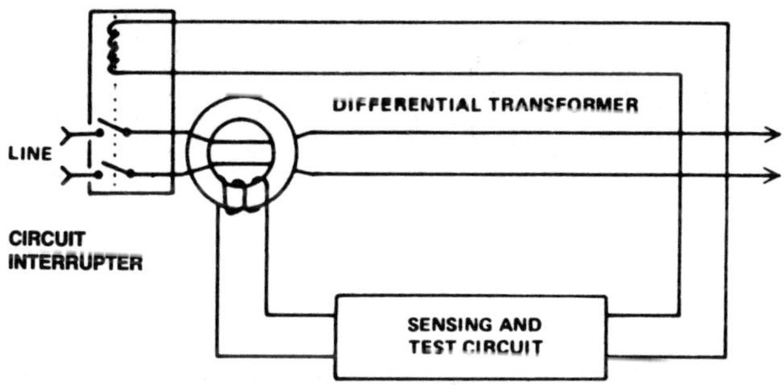

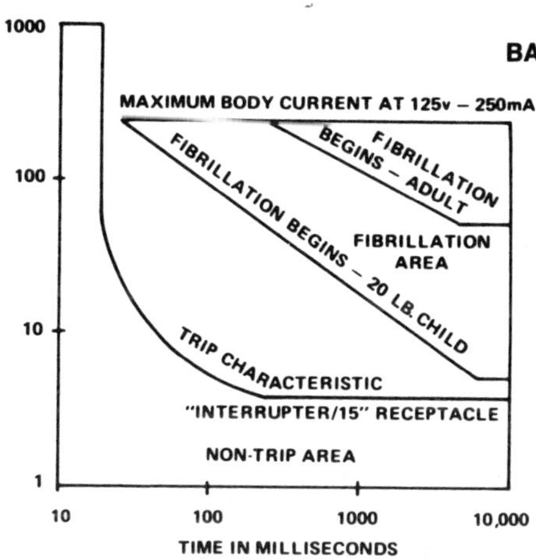

15-20 AMP RECEPTACLES IN BATHROOMS OF COMMERCIAL, INDUSTRIAL, AND ALL OTHER NONDWELLING OCCUPANCIES SHALL HAVE G.F.C.I. [SEE 210-8(b)]

FATAL FIBRILLATION OF THE HEART CAN OCCUR AFTER EXPOSURE TO AS LITTLE AS SIXTY 1/1000 AMPERE (60 milliamperes).

N.E.C. Section 210-9

Circuits Derived From Autotransformers

BRANCH CIRCUITS SHALL NOT BE SUPPLIED BY AUTOTRANSFORMERS.
EXCEPT WHERE THE SYSTEM SUPPLIED HAS A GROUNDED CONDUCTOR THAT IS ELECTRICALLY CONNECTED TO A GROUNDED CONNECTOR OF THE SYSTEM SUPPLYING THE AUTOTRANSFORMER.

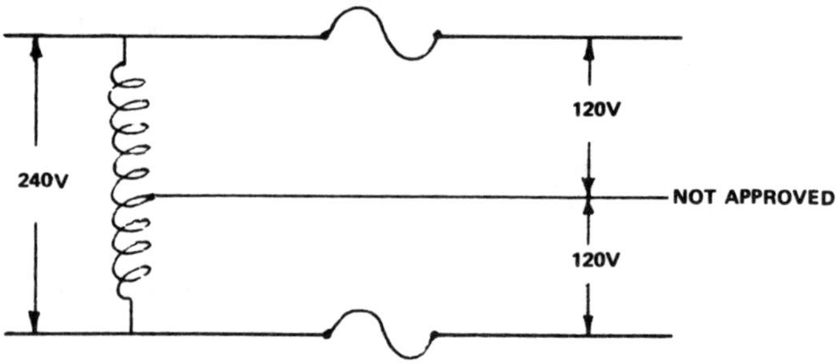

DERIVED FROM A 3Ø UNGROUNDED SYSTEM

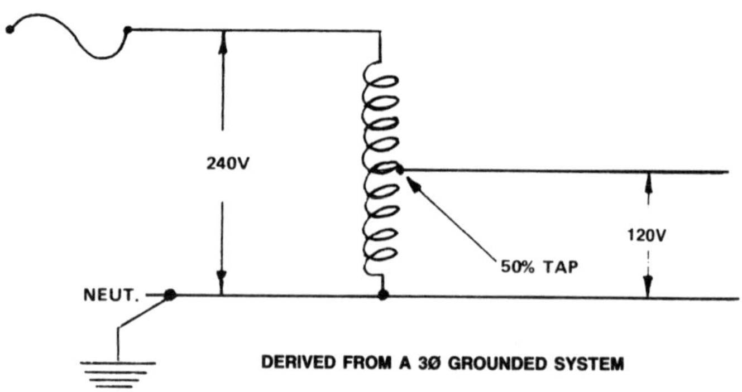

DERIVED FROM A 3Ø GROUNDED SYSTEM

FOR PROTECTION SEE 450-4

N.E.C. Section 210-19 (a) And Notes 8 And 10 To Table 310-16

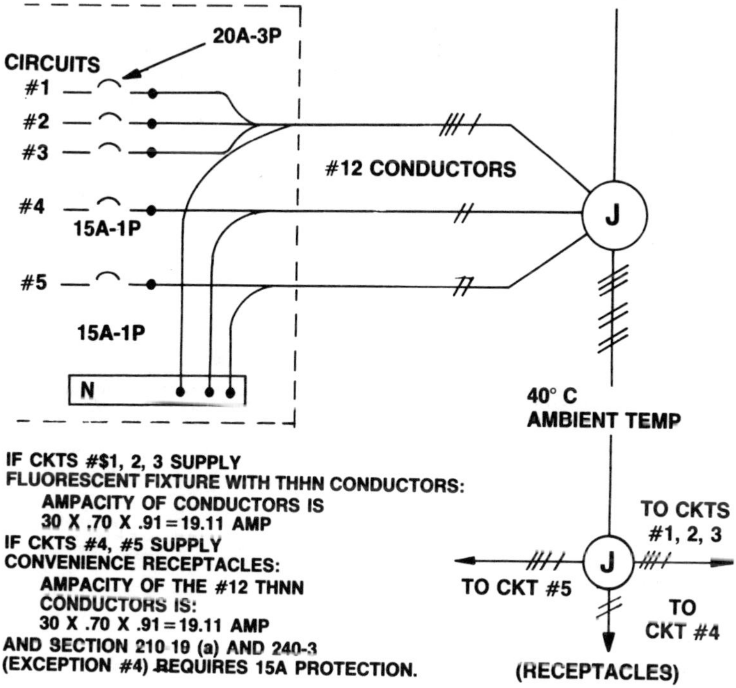

N.E.C. Sections 210-19 (a) FPN & 215-2 (b) FPN

Computing Voltage Drop, Branch Circuits

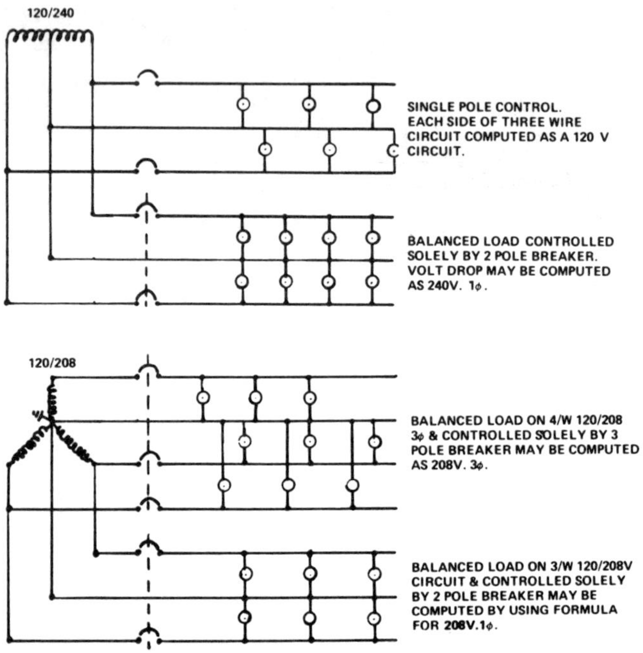

N.E.C. Sections 210-19 (a) FPN & 215-2 FPN

Branch Circuit, Feeders

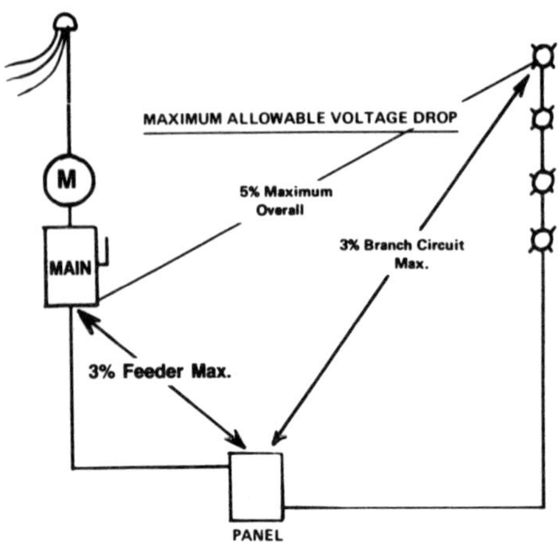

N.E.C. Section 210-19 (a)

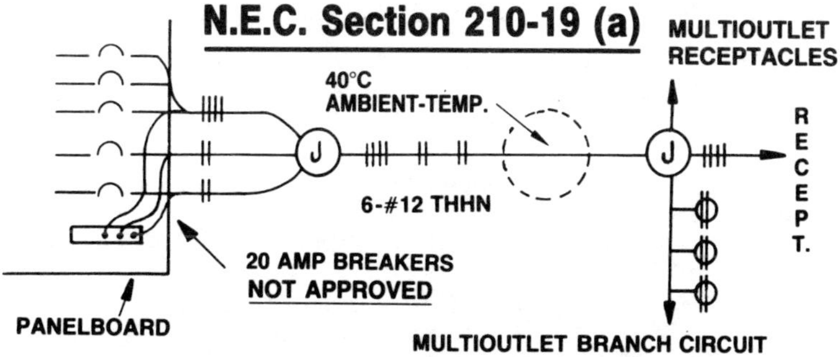

CONDUCTORS SHALL HAVE AN AMPACITY OF NOT LESS THAN THE RATING FOR CORD-AND-PLUG-CONNECTED PORTABLE LOADS.
30A × .70 × .91 = 19.11 AMP.

N.E.C. Section 210-19 (b) Ex. 2

Tap Conductors

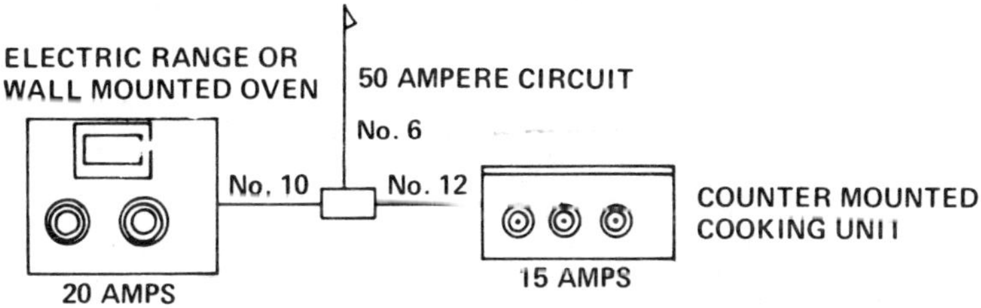

WHERE TAP CONDUCTORS SUPPLY ELECTRIC RANGES OR COOKING UNITS FROM 50-AMPERE BRANCH CIRCUITS THEY SHALL BE OF SUITABLE CAPACITY FOR THE LOAD & NOT LESS THAN 20 AMPERES IN RATING. THE TAP CONDUCTORS SHALL BE NOT LONGER THAN NECESSARY FOR SERVICING THE APPLIANCE.

N.E.C. Section 210-19 (c) Ex. 1

Tap Conductors

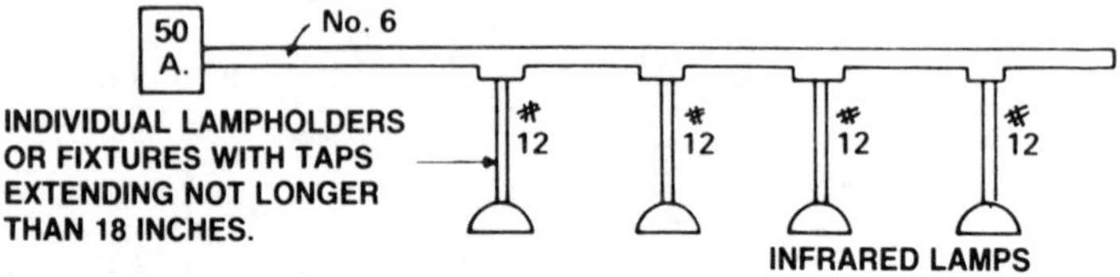

TAP CONDUCTORS MAY BE OF LESS CAPACITY THAN THE BRANCH — CIRCUIT RATING PROVIDED:

1. NO TAP CONDUCTOR IS OF LESS CAPACITY THAN THE LOAD TO BE SERVED
2. THE RATING IS NOT LESS THAN 20 AMPERES FOR 40 OR 50 AMPERE CIRCUITS OR 15 AMPERES FOR CIRCUITS RATED LESS THAN 40 AMPERES.

N.E.C. Section 210-21 (b)

Outlet Devices

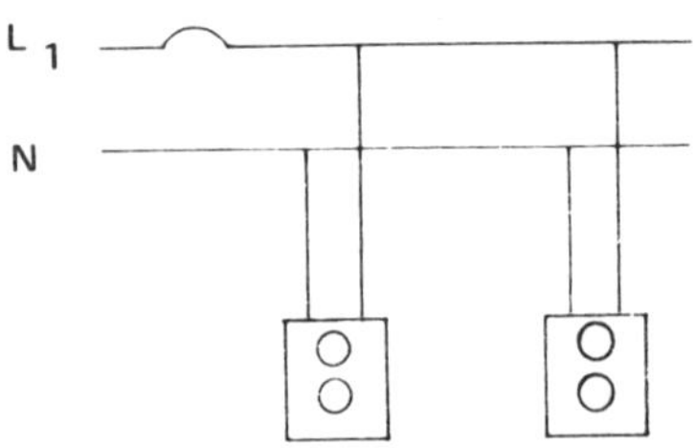

EACH 15 A RATED RECEPTACLE SHALL NOT SUPPLY AN APPLIANCE RATED MORE THAN 12 AMPERES A 20A RATED RECEPTACLE SHALL NOT SUPPLY MORE THAN 16 AMPERES.

N.E.C. Section 210-21 (b)

Receptacles Rating

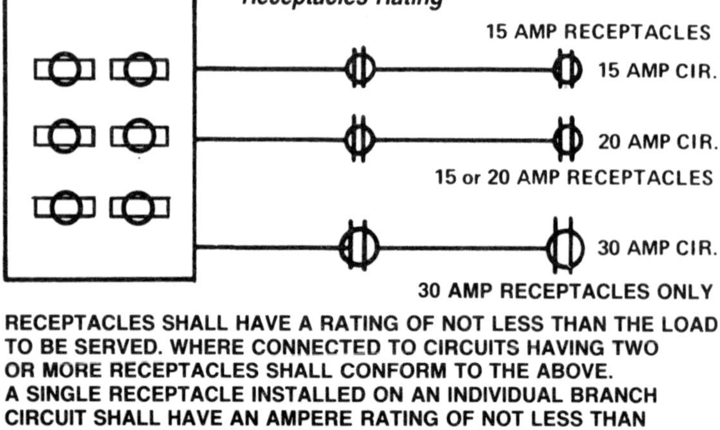

RECEPTACLES SHALL HAVE A RATING OF NOT LESS THAN THE LOAD TO BE SERVED. WHERE CONNECTED TO CIRCUITS HAVING TWO OR MORE RECEPTACLES SHALL CONFORM TO THE ABOVE. A SINGLE RECEPTACLE INSTALLED ON AN INDIVIDUAL BRANCH CIRCUIT SHALL HAVE AN AMPERE RATING OF NOT LESS THAN THAT OF THE BRANCH CIRCUIT.

N.E.C. Sections 210-21 (b) & 210-23 (b)

Portable Appliance On A 30 Ampere Circuit

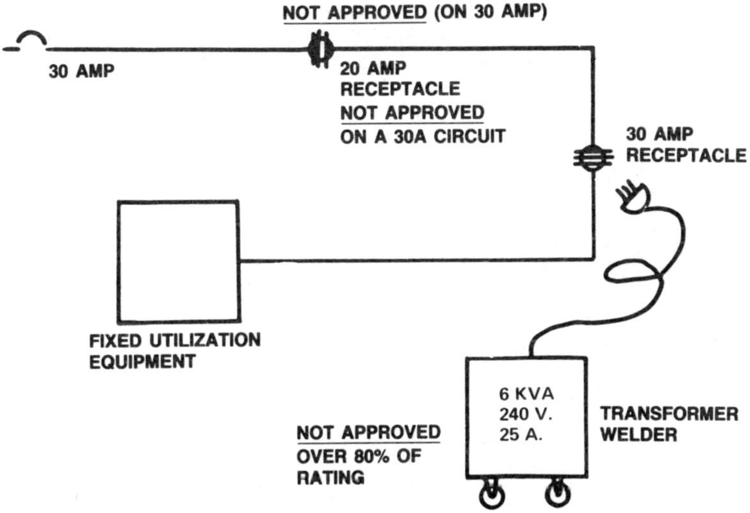

30 AMP APPLIANCE BRANCH CIRCUIT O.K. FOR 30 AMP RECEPTACLES, AND FIXED OR PORTABLE UTILIZATION EQUIPMENT.

N.E.C. Section 210-22 (c)

Branch Circuit Conductors For Continuous Loads

BRANCH CIRCUIT CONDUCTORS SUPPLYING CONTINUOUS LOADS SHALL NOT BE LOADED TO MORE THAN 80% OF THEIR CURRENT CARRYING CAPACITY.

CIRCUIT NO.	PURPOSE	OUTLETS	AMPS	MIN. CIRC. RATING
1 240 V	CAFE GRIDDLE	1	7.2 KW 30 A.	40A
2 240 V	DUCT HEATER	1	4.6 K.W. 19.1 A.	25A
3 240 V	WATER HEATER	1	3 KW 12.5 A	20A
4 120 V	OUTLINE LIGHTING CIRCUITS	6 TRANSF.	12 A.	15A
5 120 V	OUTDOOR FLOOD LIGHTS	6 LAMPS AT 500 W	3 KW 25 A.	40A

N.E.C. Section 210-22 (c)

Example Of 80% And 125% Requirements

LIMITING LOADS ON DEVICES FOR CONTINUOUS LOADS TO 80% OF THEIR RATING REQUIRES CERTAIN DEVICES TO HAVE A RATING OF NOT LESS THAN 125% OF THE CONTINUOUS LOAD RATING.

THESE REQUIREMENTS ARE THE SAME.

EXAMPLES: 40 A. LOAD (CONTINUOUS)

40 × 125% = 50 A = MIN. RATING OF O.C.P. DEVICE,
50 A. DEVICE × 80% = 40 A CONTINUOUS LOAD PERMITTED.

<u>EXCEPTION:</u> WHERE BRANCH CIRCUIT OVERCURRENT DEVICES AND ASSOCIATED EQUIPMENT ARE APPROVED FOR 100% RATING, AND ARE SO MARKED, THEY NEED NOT BE DERATED TO 80%.

Continuous Loads

CONDUCTORS SUPPLYING CONTINUOUS LOADS: CONTINUOUS LOAD INCLUDES ANY LOAD WHERE THE CURRENT IS EXPECTED TO CONTINUE FOR 3 HOURS OR MORE

COMMERCIAL BLDG	
CIRCUIT	PURPOSE
1 240 V	CAFE GRIDDLE
2 240 V	DUCT HEATER
3 240 V	WATER HEATER
4 120 V	SIGN
5 120 V	CEILING LIGHTS
6 120 V	SHOW WINDOW LIGHTS
7 120 V	SHOW CASE LIGHTS
8 120 V	FESTOON LIGHTS
9 120 V	FLOOD LIGHTS

INDUSTRIAL METAL FAB.	
CIRCUIT	PURPOSE
1 240 V	INFRA-RED DRYER
2 240 V	AIR HEATER
3 240 V	WATER HEATER
4 120 V	SHOP LIGHTS
5 120 V	OFFICE LIGHTS
6 120 V	FLOOD LIGHTS

<u>DWELLINGS</u> — AIR HEATING, DUCT BASEBOARD, CEILING, ETC. — SPACE HEATING, EXCEPT BATHROOM HEATERS. WATER HEATER.

<u>PUBLIC AREAS IN MULTIPLE APT. BUILDINGS</u> — ALL HOUSE HEATING; ALL HEATING APPLIANCES; ALL COMBINATION MOTORHEATING APPLIANCES.

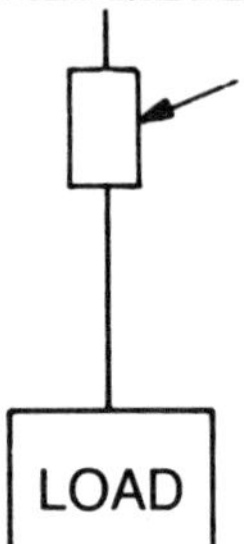

BREAKER OR FUSE MUST COMPLY WITH SEC. 240-3 AND MUST MATCH AMPACITY OF BRANCH CIRCUIT CONDUCTORS (OR BE THE NEXT HIGHER STANDARD SIZE) SEC. 240-3, EX. 1.

210-22(c) IF LOAD IS CONTINUOUS, IT CANNOT EXCEED 80 PERCENT OF THE RATING OF THE BRANCH CIRCUIT, EVEN IF THE CONDUCTORS PROTECTED HAD THEIR AMPACITY ELEVATED IN ACCORDANCE WITH NOTE 8 TO TABLES 310-16 - 310-18.

N.E.C. Section 210-23 (a)

Permissible Loads

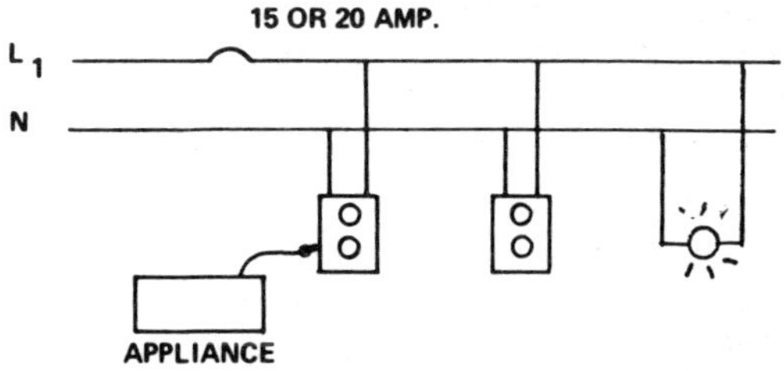

1. THE TOTAL RATING OF ANY UTILIZATION EQUIPMENT SHALL NOT EXCEED 50% OF THE BRANCH CIRCUIT RATING, WHEN LIGHTING UNITS OR RECEPTACLES ARE ALSO SUPPLIED.
2. THE RATING OF ANY CORD-AND-PLUG CONNECTED UTILIZATION EQUIPMENT SHALL NOT EXCEED 80% OF THE BRANCH CIRCUIT RATING.

N.E.C. Section 210-23

Individual Branch Circuits

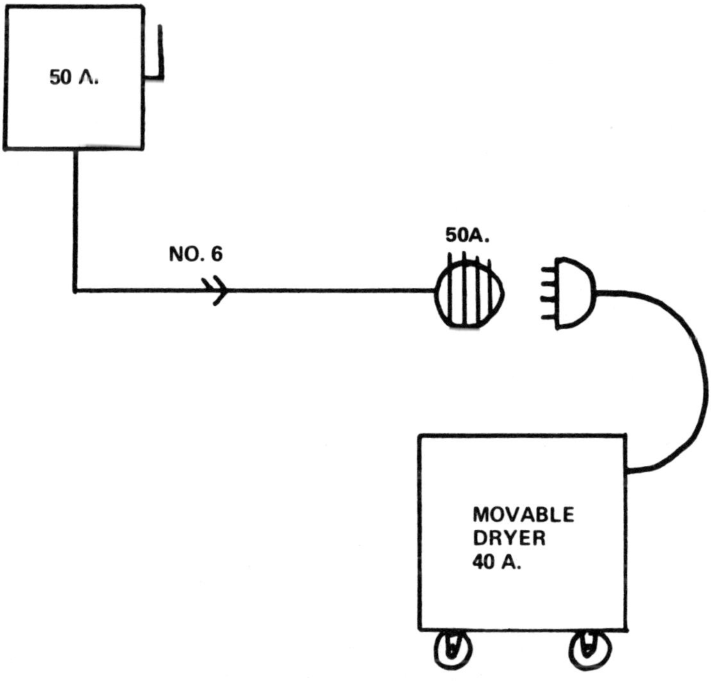

INDIVIDUAL BRANCH CIRCUIT USED TO SUPPLY A SINGLE UTILIZATION EQUIPMENT. (WHEN IN CONFORMANCE WITH ALL OTHER CODE PROVISIONS).

N.E.C. Section 210-23

Limitation Of Utilization Equipment Fastened In Place On A 20 Amp. Circuit

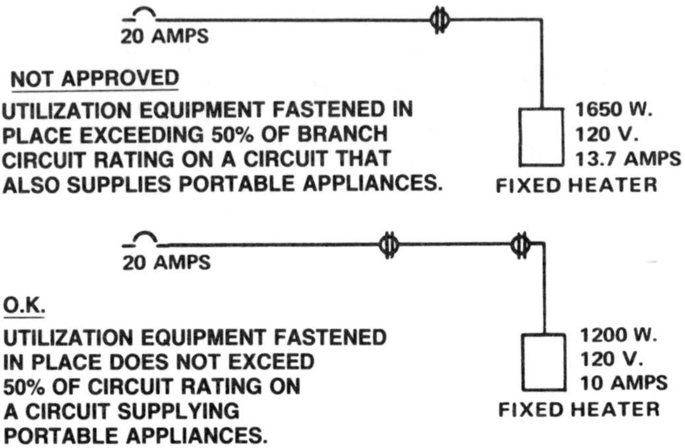

N.E.C. Section 210-23 & 210-62

Non-Dwelling Lighting

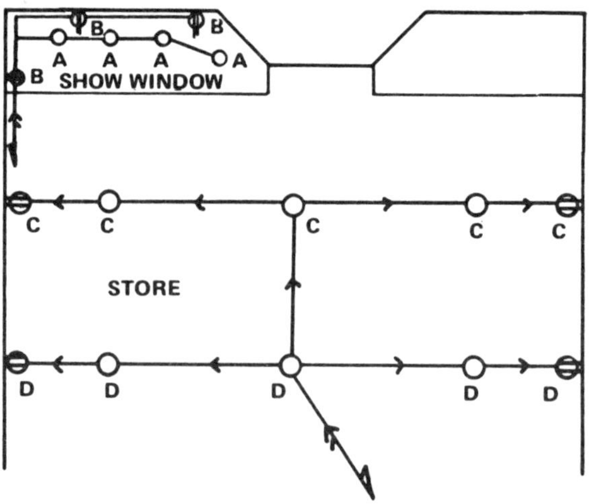

RECEPTACLES ARE PERMITTED ON GENERAL ILLUMINATION CIRCUITS

RECEPTACLES SHALL BE PROVIDED ABOVE SHOW WINDOWS FOR EACH 12 LINEAL FOOT OF SHOW WINDOW (210-62).

N.E.C. Section 210-23 (a)

Permissible Load On A 20 Amp. Circuit

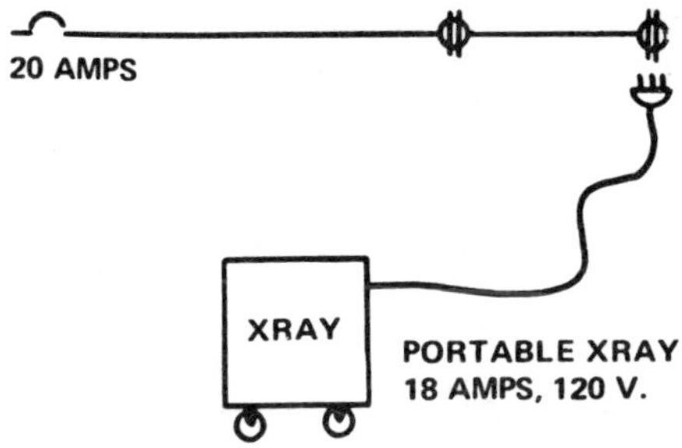

NOT APPROVED
RATING OF PORTABLE APPLIANCE (ON CIRCUIT WITH OTHER LOAD) EXCEEDS 80% OF CIRCUIT RATING.

N.E.C. Section 210-23 (a) (b) (c) (d)

Ampere Rating Of Lighting Circuits (Non-Dwelling)

15A
OR
20A

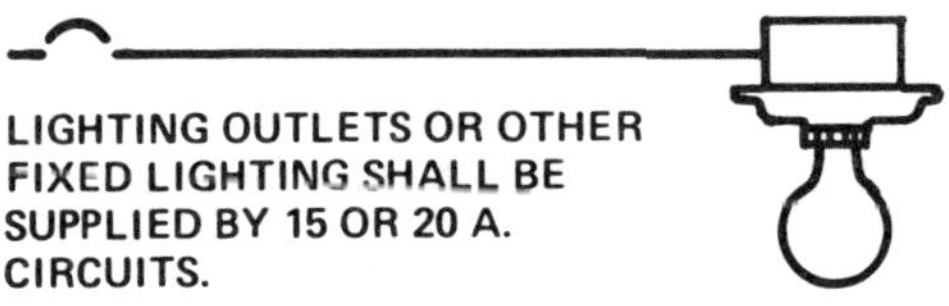

LIGHTING OUTLETS OR OTHER FIXED LIGHTING SHALL BE SUPPLIED BY 15 OR 20 A. CIRCUITS.

EXCEPTION:

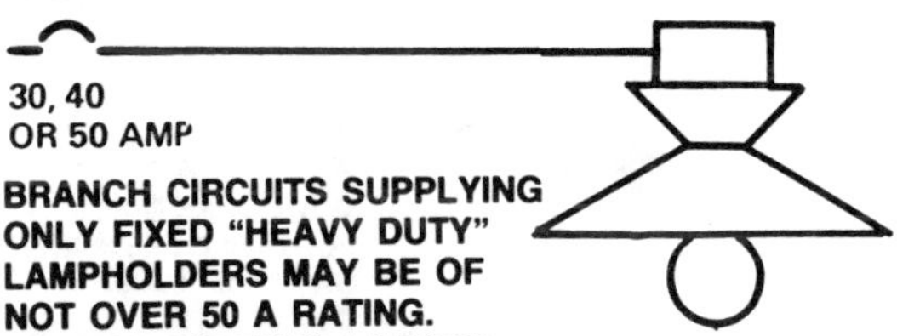

30, 40
OR 50 AMP

BRANCH CIRCUITS SUPPLYING ONLY FIXED "HEAVY DUTY" LAMPHOLDERS MAY BE OF NOT OVER 50 A RATING. BRANCH CIRCUITS LARGER THAN 50 A SHALL SUPPLY ONLY NONLIGHTING OUTLET LOADS.

N.E.C. Section 210-23 (b)

Thirty Ampere Branch Circuit

THIRTY AMPERE APPLIANCE BRANCH CIRCUITS MAY ONLY BE USED TO SUPPLY RECEPTACLE OUTLETS EQUIPPED WITH 30 AMP RECEPTACLES OR APPLIANCES, FIXED OR PORTABLE OR, FIXED LIGHTING UNITS WITH HEAVY DUTY LAMP HOLDERS IN OTHER THAN DWELLING UNITS BUT NOT A COMBINATION OF BOTH.

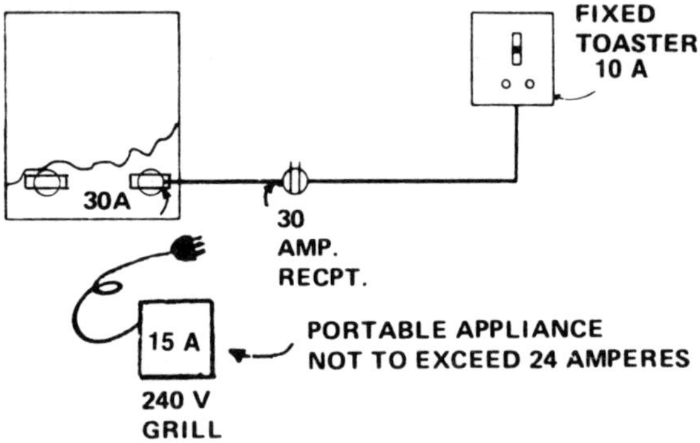

N.E.C. Section 210-23 (c)

40 And 50 Ampere Appliance Branch Circuits

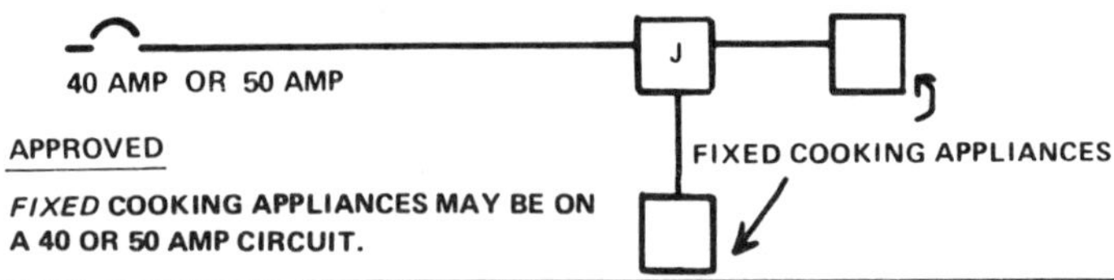

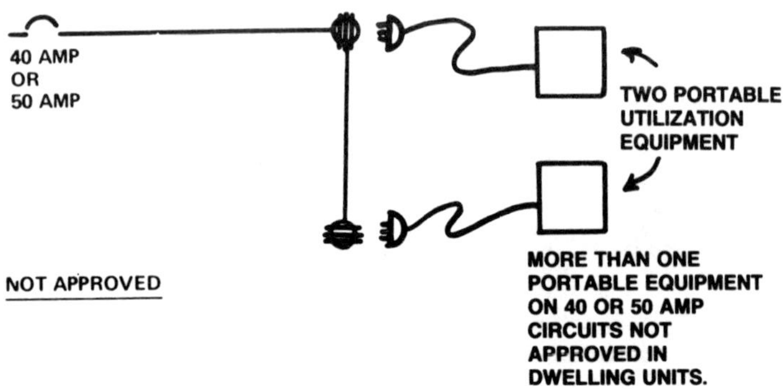

N.E.C. Section 210-24, Exception

BRANCH CIRCUITS IN DWELLING UNITS SHALL NOT BE CONNECTED TO SERVE MORE THAN UNIT

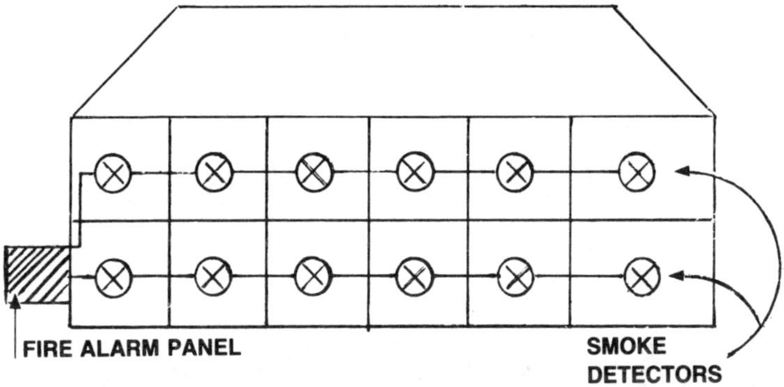

THIS EXCEPTION APPLIES TO CIRCUITS IN TWO FAMILY OR MULTIFAMILY DWELLINGS UNDER SINGLE MANAGEMENT, WHERE BRANCH CIRCUITS SUPPLY COMMUNICATIONS, FIRE ALARMS, SIGNAL OR SIMILAR FUNCTIONS.

N.E.C. Section 210-52

Location Of Receptacles Exceptions

GENERAL USE RECEPTACLES WHERE REQUIRED BY THE CODE SHALL BE INSTALLED.

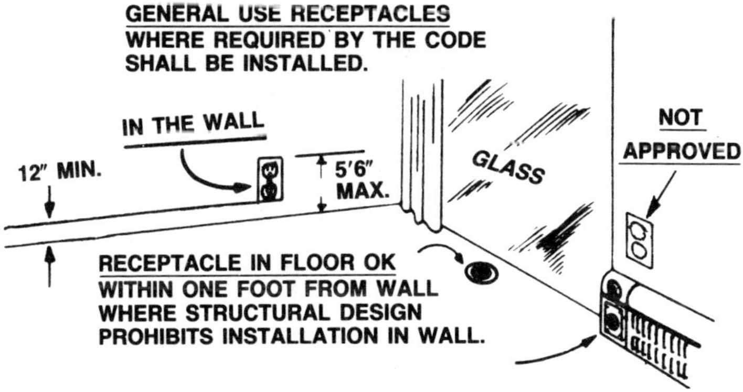

RECEPTACLES IN BASEBOARD SHALL NOT BE CONNECTED TO HEATER CIRCUIT

N.E.C. Section 210-52

Receptacles In Kitchen & Dining Rooms & 210-8 (a) 5

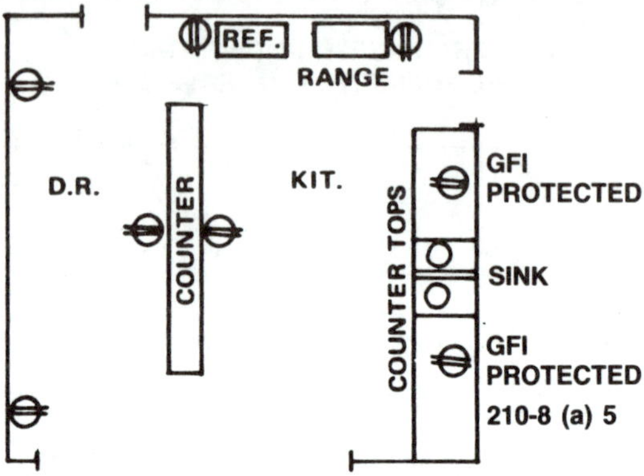

1. A RECEPTACLE OUTLET SHALL BE INSTALLED SO THAT NO POINT ALONG THE WALL SPACE IS MORE THAN 6 FEET FROM A RECEPTACLE OUTLET.
2. A RECEPTACLE SHALL BE INSTALLED AT EACH COUNTER SPACE WIDER THAN 12".

NOTE–1. COUNTER SPACES SEPARATED BY SINKS, RANGE TOPS OR REFRIGERATORS ARE CONSIDERED AS SEPARATE COUNTER TOP SPACES.

NOTE–2. RECEPTACLES RENDERED INACCESSIBLE BY INSTALLATION OF STATIONARY APPLIANCES (REFRIGERATOR) WILL NOT BE CONSIDERED AS THESE REQUIRED RECEPTACLES.

N.E.C. Section 210-52 (a)

Spacing Of Receptacles

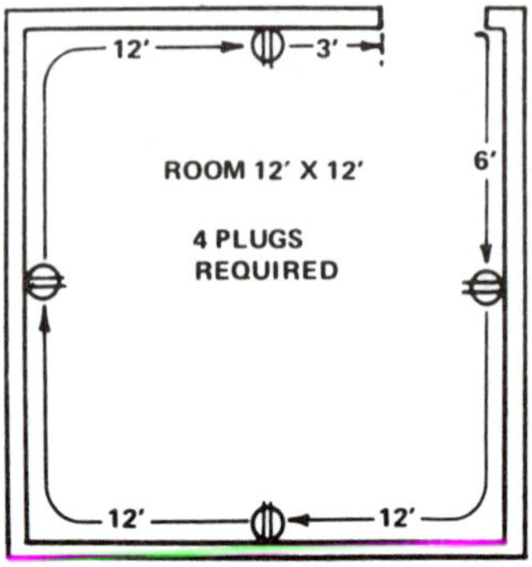

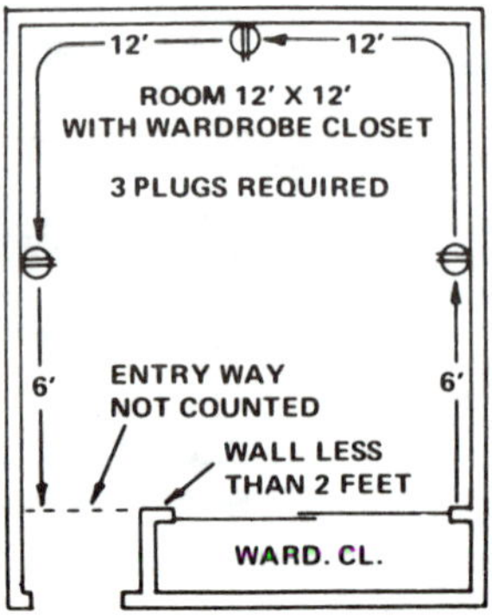

N.E.C. Section 210-52 (c)

Location Of Receptacles In Kitchen

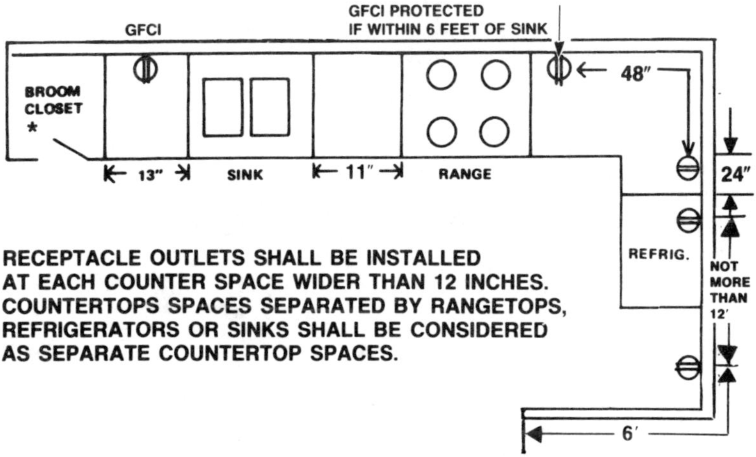

RECEPTACLE OUTLETS SHALL BE INSTALLED AT EACH COUNTER SPACE WIDER THAN 12 INCHES. COUNTERTOPS SPACES SEPARATED BY RANGETOPS, REFRIGERATORS OR SINKS SHALL BE CONSIDERED AS SEPARATE COUNTERTOP SPACES.

N.E.C. Section 210-52 (H)

Hallways Over 10′ In Length Require One Receptable

HALLWAYS 10 FEET OR LONGER SHALL HAVE AT LEAST ONE RECEPTACLE.

N.E.C. Sections 210-52 (b) & 210-8 (a) 5

Location Of Receptacles In Kitchen

RECEPT. OUTLETS REQ. IN EACH COUNTER SPACE OVER 12″ WIDE.

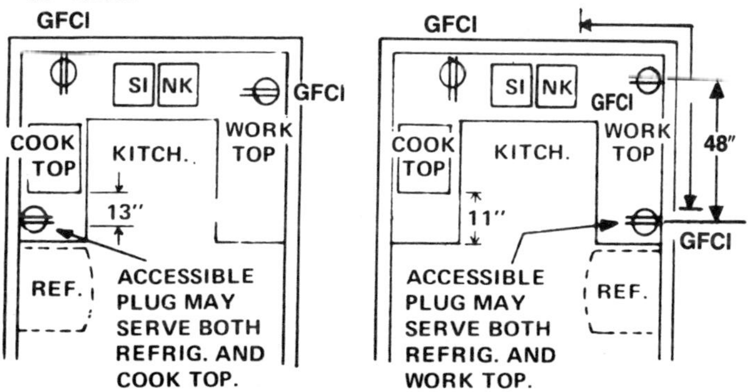

COUNTER SPACES SEPARATED BY RANGETOPS, REFRIG. OR SINKS SHALL BE CONSIDERED A SEPARATE COUNTERTOP SPACE.

N.E.C. Section 210-60 & 210-8 (b)

Receptacles In Guest Room

RECEPTACLE OUTLETS SHALL BE SPACED AS REQUIRED FOR DWELLINGS AS SPECIFIED IN SECTION 210-52.

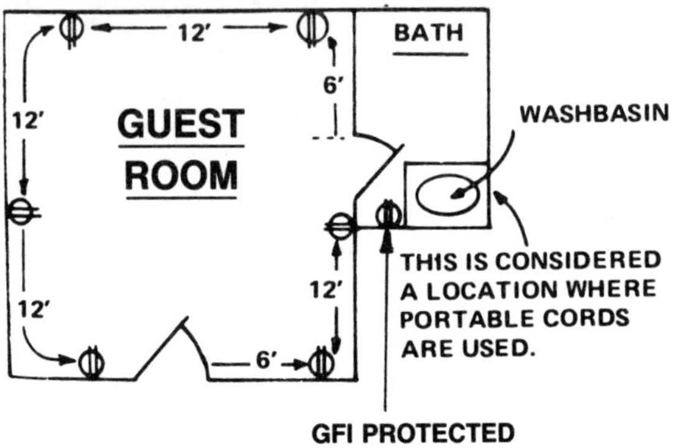

EXCEPTION: IN GUEST ROOMS THE REQUIRED NUMBER OF RECEPTACLE OUTLETS SHALL BE PERMITTED TO BE LOCATED FOR PERMANENT FURNITURE LAYOUT.

N.E.C. Section 210-60 Exception & 210-8 (b)

Receptacles Outlets Required

GUEST ROOM IN HOTEL AND MOTEL

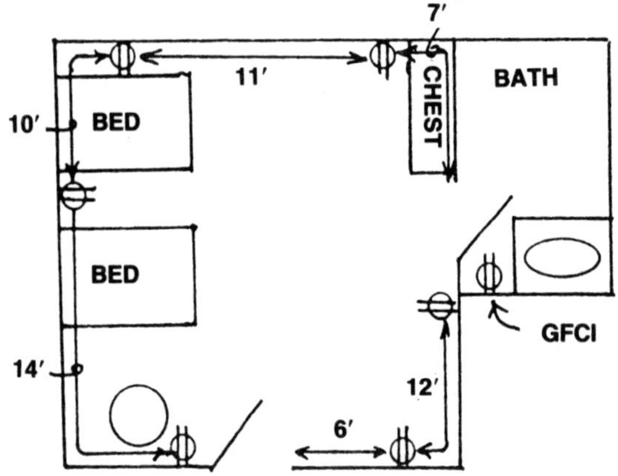

RECEPTACLES MAY BE LOCATED CONVENIENT TO THE PERMANENT FURNITURE LAYOUT. THE TOTAL NUMBER OF RECEPTACLES SHALL NOT BE REDUCED TO LESS THAN ONE RECEPTACLE FOR EACH 12' LINEAL FEET OF WALL SPACE OR FRACTION THEREOF.

N.E.C. Section 210-62

Show Window

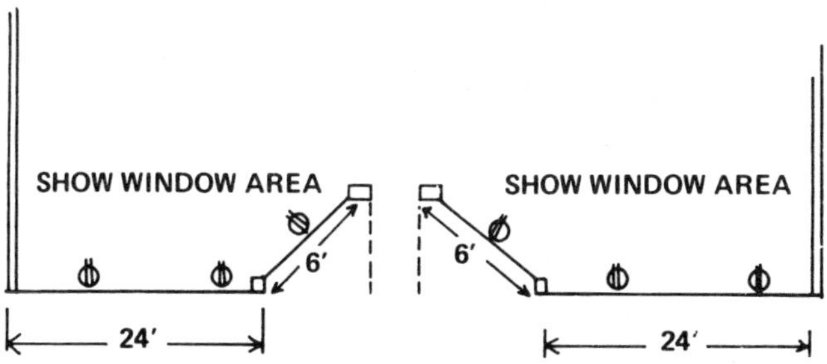

AT LEAST ONE RECEPTACLE SHALL BE PROVIDED DIRECTLY ABOVE A SHOW WINDOW FOR EACH 12 LINEAL FEET OR MAJOR FRACTION THEREOF. THIS REQUIREMENT IS IN ADDITION TO THE ILLUMINATION CIRCUITS THAT ARE REQUIRED TO BE INSTALLED.

N.E.C. Section 210-63

RECEPTACLE REQUIRED FOR MAINTENANCE OF MECHANICAL EQUIPMENT.

EXCEPTION: ROOFTOP EQUIPMENT ON ONE AND TWO FAMILY DWELLINGS.

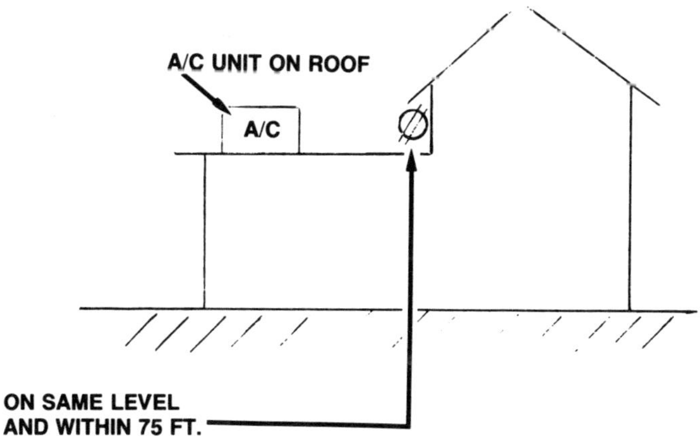

NOTE: CALIF. MECHANICAL CODE AND ELECTRICAL CODE REQUIRE A RECEPTACLE WITHIN 25 FT.

N.E.C. Section 210-70 (a)

Required Lighting Outlets Dwellings

AT LEAST ONE WALL SWITCH-CONTROLLED LIGHTING OUTLET SHALL BE INSTALLED IN EVERY HABITABLE ROOM; IN BATHROOMS, HALLWAYS, STAIRWAYS, ATTACHED GARAGES AND DETACHED GARAGES WITH ELECTRIC POWER; AND AT OUTDOOR ENTRANCES OR EXITS.

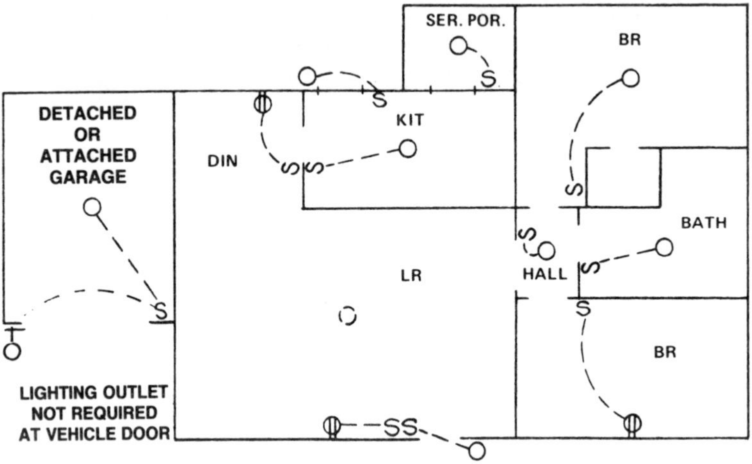

N.E.C. Section 210-70 (a)

Lighting Outlets Dwellings

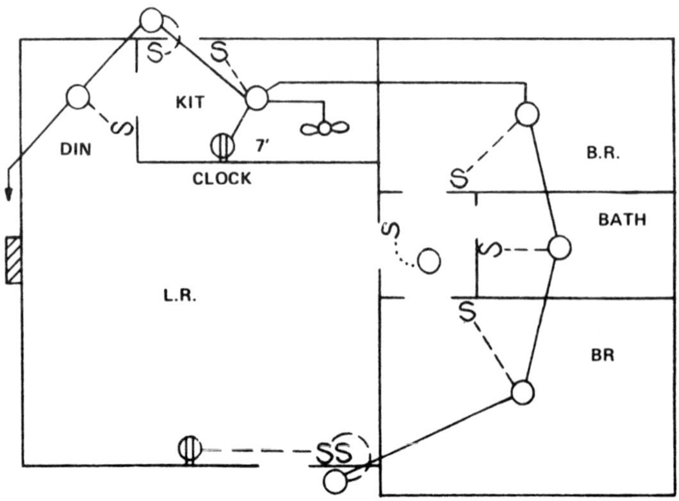

N.E.C. Section 215-2

Minimum Feeder Sizes [SEE 210-19 (a)]

MINIMUM SIZE FEEDERS		
CONDITION	*DESCRIPTION*	*SIZE*
GENERAL	FEEDER CONDUCTORS SHALL HAVE AN AMPACITY NOT LOWER THAN REQUIRED TO SUPPLY THE LOAD AS COMPUTED IN PARTS B, C AND D OF ARTICLE 220.	
SPECIFIC	TO CARRY THE TOTAL CURRENT SUPPLIED BY A SERVICE.	$\leq$ 55 A.
	TO SUPPLY 2 OR MORE BRANCH CIRCUITS.	MIN. 30 A.

N.E.C. Section 215-2

Feeder Conductors

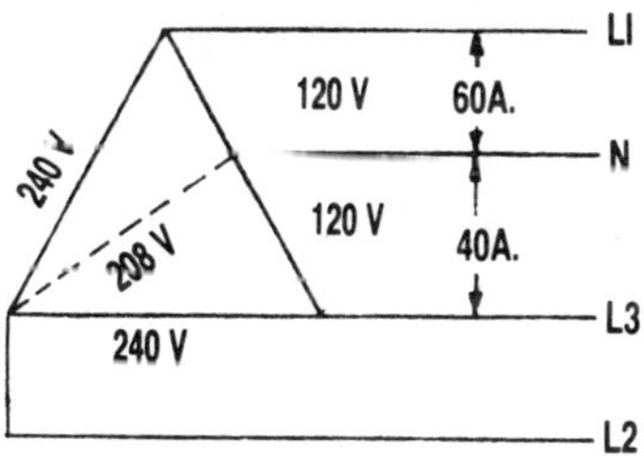

METHOD FOR COMPUTING 3ø 4W 120/240V.
DELTA SYSTEM
A BUILDING HAVING A 3ø LOAD OF 30 AMPS
240V AND A SINGLE PHASE LOAD OF 100 AMPS
WITH 60 AMPS LINE TO NEUTRAL & 40 AMPS
LINE TO NEUTRAL.

ADD HIGHEST NEUTRAL TO LINE LOAD TO THE TOTAL 3ø LOAD

NEUTRAL TO LINE	60A	
3ø	30A	FEEDER WITH REDUCED NEUTRAL
	90A	AS PERMITTED BY 220-22

N.E.C. Section 215-10

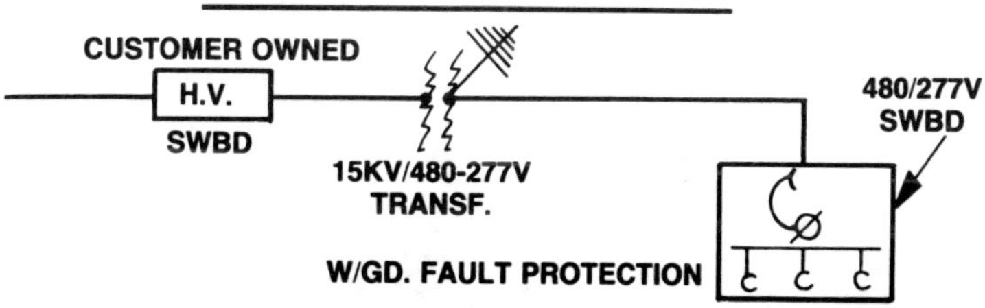

N.E.C. Section 215-2 (a)

Minimum Feeder Sizes

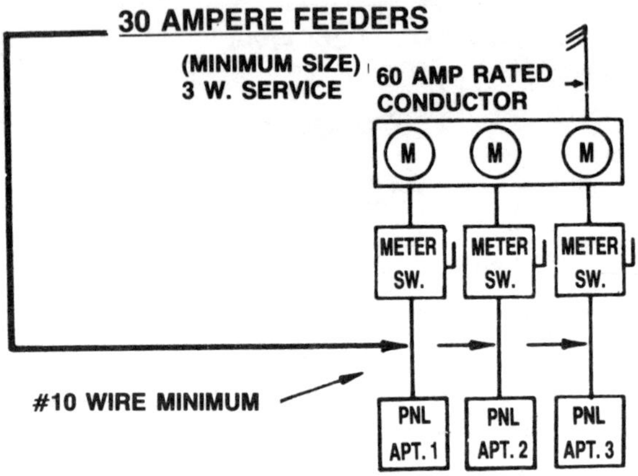

N.E.C. Sections 215-2 (b) & 230-42

Dwellings — Minimum Feeder Sizes

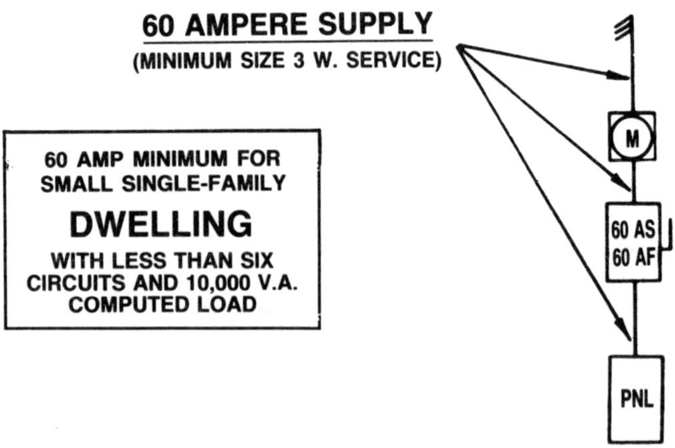

THE FEEDER CONDUCTORS SHALL NOT BE SMALLER THAN SERVICE CONDUCTOR WHERE THEY CARRY ALL THE LOAD OF THE SERVICE.

N.E.C. Section 215-4

Common Neutral Feeder

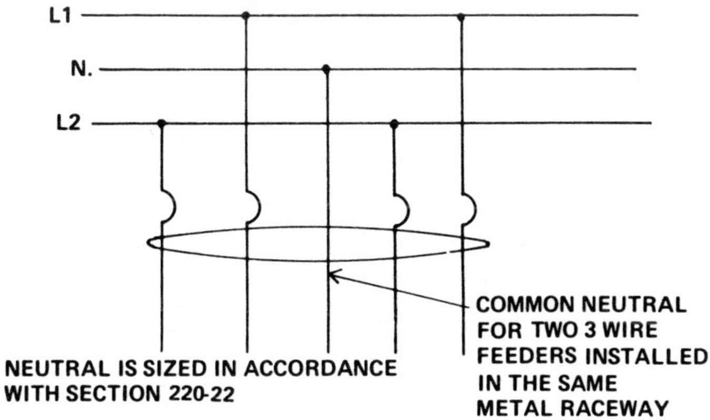

N.E.C. Section 215-10 (& 230-95)

GROUNDED WYE SYSTEM, GREATER THAN 150 VOLTS PHASE-TO-GROUND, AND NO MORE THAN 600 VOLTS PHASE-TO-PHASE, DISCONNECTS OF 1000 AMPS OR MORE.

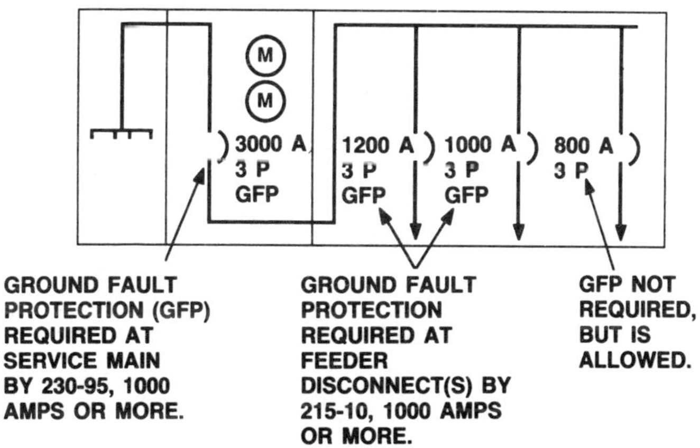

ALL GFP DEVICES SHOULD BE COORDINATED FOR PROPER SELECTIVE FUNCTION. SEE EXCEPTIONS IN 230-95(a). THE GFP PROTECTIVE SYSTEM MUST BE FIELD TESTED.

N.E.C. Section 220-3 (b)

Minimum Computed Loads On A Feeder, By Occupancies

THE FOLLOWING EXAMPLES SHOW THE MINIMUM COMPUTED WATTAGES FOR DIFFERENT GENERAL ILLUMINATION LOADS ON A *FEEDER* SERVING A GARAGE AREA. SEE TABLE 220-3(b) FOR OTHER VALUES.

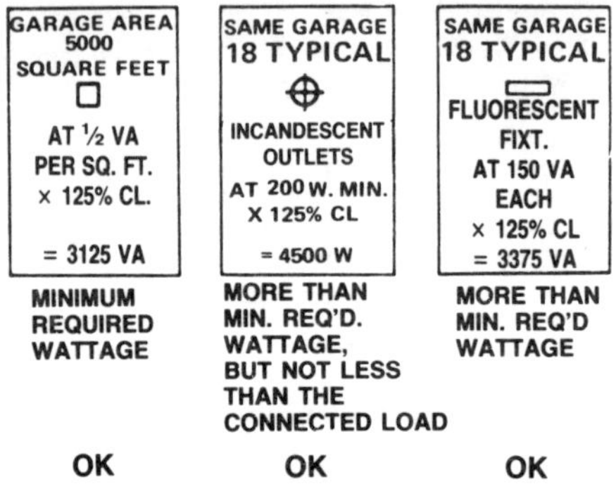

N.E.C. Section 220-3 (c) & 210-23

Minimum Load Rating On Branch Circuits

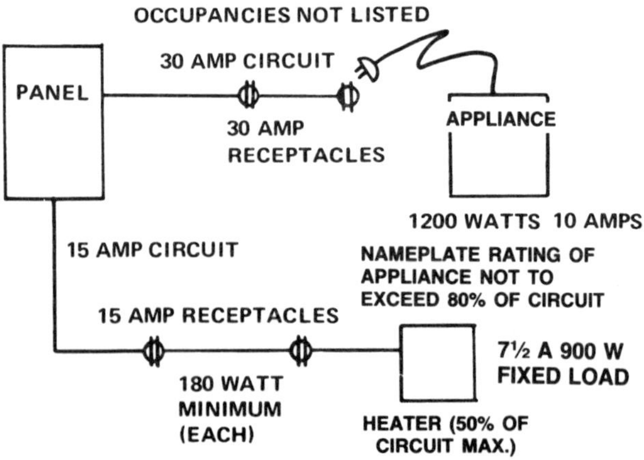

COMPUTING LOAD OF APPLIANCE AND RECEPTACLE CIRCUIT.

180 WATTS MINIMUM FOR RECEPTACLES OR RATING OF APPLIANCES (WHICHEVER IS GREATER)

N.E.C. Sections 220-3 & 220-4

Minimum Number Of Lighting Circuits Required

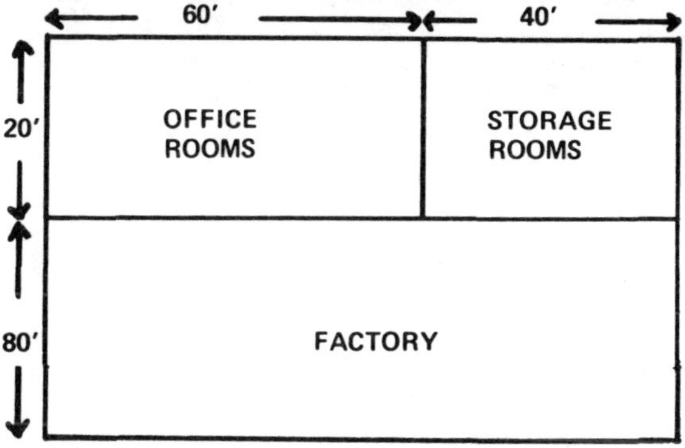

EXAMPLE'S VOLTAGE = 120

*OFFICE	20′ X 60′ X 3½ VA/SQ. FT. =	4,200
STORAGE	20′ X 40′ X ¼ VA/SQ. FT. =	200
FACTORY	100′ X 80′ X 2 VA/SQ. FT. =	1,600
	VA TOTAL =	20,400 VA
10 RECEPTACLES PLUS 25% L.C.L.	=	5,100 VA
		25,500 VA

$$\frac{\text{25,500 VA REQUIRED}}{\text{(120 X 20) VA PER CIRCUIT}} = \text{10.6 CIRCUITS} = \text{11 CIRCUITS MINIMUM.}$$

Table 220-3(b). General Lighting Loads by Occupancies

Type of Occupancy	Unit Load per Sq. Ft. (Volt-Amperes)
Armories and Auditoriums	1
Banks	3½**
Barber Shops and Beauty Parlors	3
Churches	1
Clubs	2
Court Rooms	2
*Dwelling Units	3
Garages — Commercial (storage)	½
Hospitals	2
*Hotels and Motels, including apartment houses without provisions for cooking by tenants	2
Industrial Commercial (Loft) Buildings	2
Lodge Rooms	1½
Office Buildings	3½**
Restaurants	2
Schools	3
Stores	3
Warehouses (storage)	¼
In any of the above occupancies except one-family dwellings and individual dwelling units of two-family and multifamily dwellings:	
Assembly Halls and Auditoriums	1
Halls, Corridors, Closets, Stairways	½
Storage Spaces	¼

For SI units: one square foot = 0.093 square meter.

* All general use receptacle outlets of 20-ampere or less rating in one-family, two-family and multifamily dwellings and in guest rooms of hotels and motels [except those connected to the receptacle circuits specified in Sections 220-4(b) and (c)] shall be considered as outlets for general illumination, and no additional load calculations shall be required for such outlets.

** In addition a unit load of 1 volt-ampere per square foot shall be included for general purpose receptacle outlets when the actual number of general purpose receptacle outlets is unknown.

NOTE: *The above Table 220-3(b) is a direct reproduction from the 1991 National Electrical Code. It is reproduced here as handy reference on the subject of feeders (Article 220 of the N.E.C.) as covered in pages immediately ensuing.*

N.E.C. Section 220-3, 10, 11, 12

Computation Of Feeder Load — Non-Dwelling

COMPUTING FEEDER LOADS FOR STORES

STORE BLDG. 120/240 V 1ϕ	WATTS	TOTAL
120 LIGHT OUTLETS @220W EA	24,000	
85 BALLASTS 1.7 AMPS @ 120 V	17,340	
22½ FT SHOW WINDOW 200 W per foot	4500	
1 SIGN	1200	
CONTINUOUS LOAD 125% x	47,040	58,800
56 GEN. USE RECEPT. 180 W. 100%	**10,080**	**10,080**
	57,120	**68,880**
57,120 = 238A 68,880 = 287A		

a) THIS REQUIRES 250 MCM, THW CONDUCTORS BUT 220-10(b) REQUIRES THAT THE FEEDER OVERCURRENT MUST BE RATED NOT LESS THAN THE TOTAL LOAD INCLUDING 125% OF THE CONTINUOUS LOAD WHEN USING A DEVICE THAT IS NOT UL-LISTED FOR CONTINUOUS OPERATION AT 100% OF RATING.

b) THE FEEDER PROTECTIVE DEVICE SHOULD THEN BE $\frac{68{,}880\text{ VA}}{240\text{V}} = 287$ A OR 300A.

c) ALTHOUGH THE FEEDER CONDUCTORS WITH AN AMPACITY OF 255A ARE ADEQUATE TO CARRY 238A CONTINUOUSLY, THEY WOULD NOT BE PROPERLY PROTECTED BY A 300A BREAKER (SEE SECT. 240-3).

d) ACCORDING TO TABLE 310-6, 300 MCM CONDUCTORS MUST BE USED TO MATCH THE 300A OVERCURRENT DEVICE.

N.E.C. Section 220-3

General Illumination Feeder Load

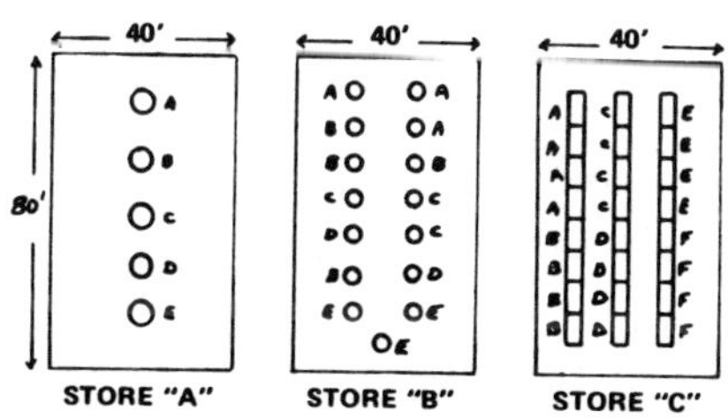

MINIMUM WATTAGE EACH STORE
40′ X 80′ X 3 W/SQ FT 9600 WATTS

STORE "A" 5 OUTLETS X 1920 W = 9600 = 5 CIR.
STORE "B" 15 OUTLETS X 640 W = 9600 = 5 CIR.
STORE "C" 48 BALLASTS
1.9 AMPS X 120 VOLTS = 10,944 = 6 CIR.

N.E.C. Section 220-3 (c)

Maximum Number Of Receptacles On A 20 Ampere Circuit

LOAD ON BRANCH CIRCUITS NON-DWELLING
MAXIMUM NUMBER OF OUTLETS ON 20 AMP.
120 V. RECEPTACLE CIRCUITS.

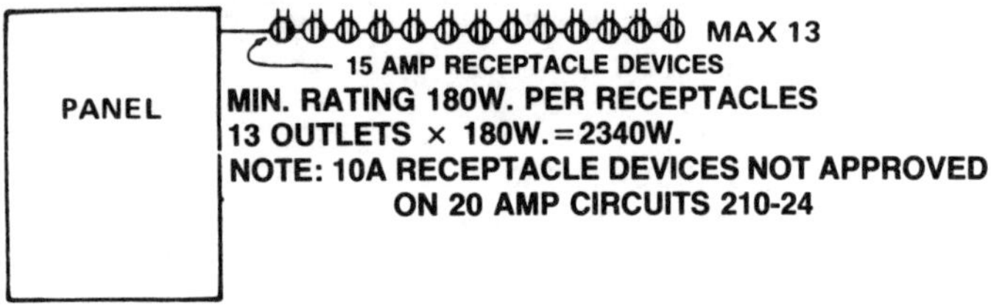

THE LOAD ON A SINGLE RECEPTACLE THAT SUPPLIES A *REFRIGERATION TYPE AIR CONDITIONING UNIT* SHALL BE COMPUTED AT NOT LESS THAN THE *NAME-PLATE RATING* OF THE APPLIANCE.

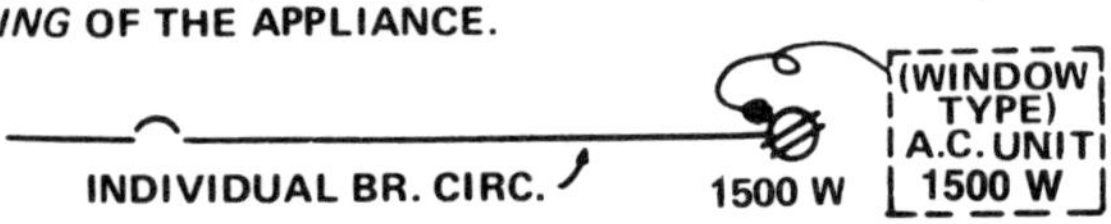

COMPUTED AT NOT LESS THAN 1500 WATTS

N.E.C. Section 220-3 (c) 5

VA Of General Use Receptacles

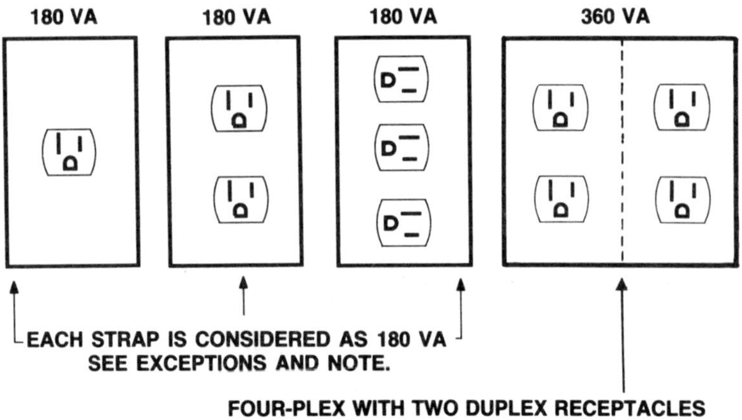

N.E.C. Section 220-3, 11, 12 & 13

Computation Of Feeder Load Non-Dwelling

COMPUTING FEEDER LOAD FOR OFFICE BLDG .

MINIMUM WATTAGE	VA	DEMAND TOTAL
1800 SQ. FT. OFFICE AREA X 3½ PER SQ. FT.	63,000	
SIGN	1,200	
3000 SQ. FT. HALLS X ½ VA/SQ. FT.	1,500	65,700
CONTINUOUS LOAD X 25%	16,425	16,425
RECEPTACLES 1 WATT PER SQ. FT.	1,800	1,800
MIN. WATTAGE		83,925

NOTE: GENERAL USE RECEPTACLE AND OTHER LOADS TO BE ADDED.

NOTE: WHERE BRANCH CIRCUIT OVERCURRENT DEVICES AND ASSOCIATED EQUIPMENT ARE APPROVED FOR 100% RATING, AND ARE SO MARKED, THEY NEED NOT BE DERATED TO 80%.

COMPUTING FEEDER LOAD FOR STORES

STORE BUILDING 120/240 V 1 ø		
	VA	TOTAL
*120 LIGHT OUTLETS, @ 200 W EACH	24000	
85 BALLASTS 1.7 AMPS	17340	
22½′ SHOW WINDOW, 200 W Per Ft.	4500	
1 SIGN CIRCUIT	1200	
CONT. LOAD X 125%	47,040 =	58,800
60 X 180 = 10,800 VA 60 GEN. USE REC. 180 VA	50% DEMAND OVER 10 KVA	10,400
TOTAL		69,200

$\frac{69{,}200}{240}$ = 288+ AMPS

NOTE: WHERE BRANCH CIRCUIT OVERCURRENT DEVICES AND ASSOCIATED EQUIPMENT ARE APPROVED FOR 100% RATING, AND SO MARKED, THEY NEED NOT BE DERATED TO 80%.

FOR GENERAL LIGHTING LOAD FEEDER DEMAND FACTORS SEE 220-11

*MINIMUM 3 WATTS PER SQ. FT.

N.E.C. Sections 220-3 (b), 220-4 (d)

Lighting Load. Branch Circuits Required

THE UNIT LOAD DWELLING IS BASED ON THE SQ. FOOT OF THE DWELLING. THERE IS NO LIMIT AS TO THE NUMBER OF OUTLETS INSTALLED ON SMALL APPLIANCE OR GENERAL ILLUMINATION BRANCH CIRCUITS.

THE NUMBER OF OUTLETS ON THE GENERAL ILLUMINATION BRANCH CIRCUITS SHALL BE EVENLY PROPORTIONED AMONG THE BRANCH CIRCUITS.

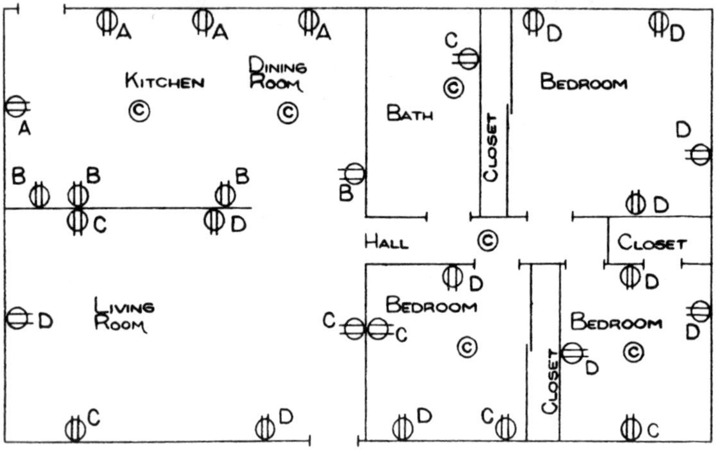

EXAMPLE:	LOAD
CIR A – 4 OUTLETS SMALL APPLIANCE	= 1500 WATTS
CIR B – 4 OUTLETS SMALL APPLIANCE	= 1500 WATTS
CIR C – 12 OUTLETS GENERAL ILLUMINATION	= 3 WATTS PER SQ FT
CIR D – 12 OUTLETS GENERAL ILLUMINATION	= 3 WATTS PER SQ FT

N.E.C. Section 220-4 (b) (1) & (2)

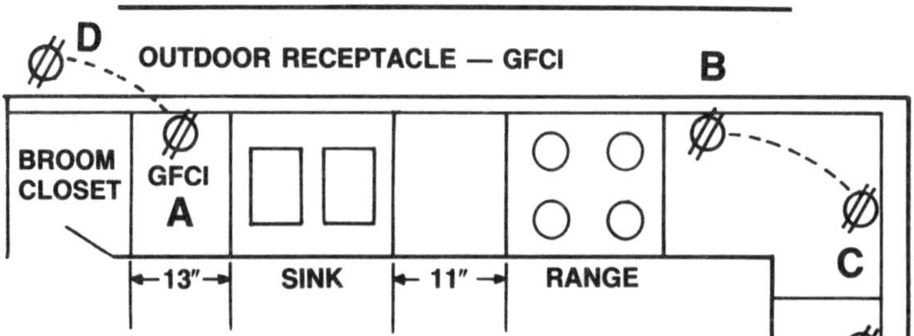

KITCHEN COUNTER RECEPTACLES MAY NOT ALL BE SUPPLIED BY THE SAME TWO WIRE SMALL APPLIANCE BRANCH CIRCUIT.

RECEPTACLES A & B MUST BE ON DIFFERENT CIRCUITS. RECEPTACLE C (CLOCK RECEPTACLE) MAY BE CONNECTED TO A SMALL APPLIANCE BRANCH CIRCUIT, EXC. #1. OUTDOOR RECEPTACLE D MAY BE SUPPLIED BY A SMALL APPLIANCE BRANCH CIRCUIT, EXC. #2.

N.E.C. Section 220-4 (b)

Appliance Branch Circuits

NOT LESS THAN TWO APPLIANCE CIRCUITS ARE REQUIRED TO SUPPLY RECEPTACLES IN THE KITCHEN.

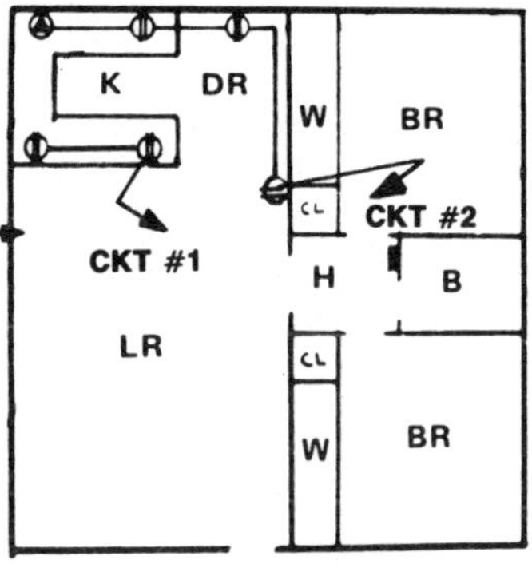

NO DISHWASHER, DISPOSER, ETC. ON SMALL APPLIANCE CIRCUITS.

N.E.C. Section 220-4 (b) & 210-52 (b) (1), Exc. #3

Appliance Branch Circuits

CIRCUITS SUPPLYING RECEPTACLES IN A SMALL APPLIANCE AREA ARE CLASSIFIED AS SMALL APPLIANCE CIRCUITS. AT LEAST TWO APPLIANCE CIRCUITS ARE REQUIRED IN THE KITCHEN.

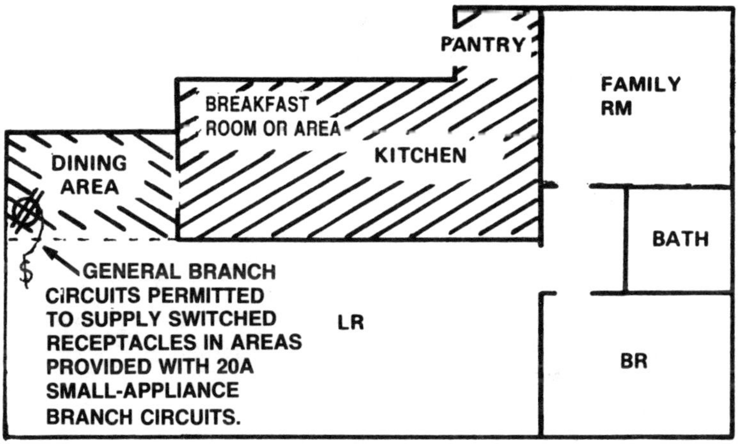

SHADED PORTIONS SHOW TYPICAL SMALL APPLIANCE AREAS

N.E.C. Section 220-10, 11, 12

Minimum Feeder Sizes — Non-Dwelling

FEEDERS SHALL BE SIZED FOR *NOT LESS* THAN MINIMUM REQUIRED WATTAGE.

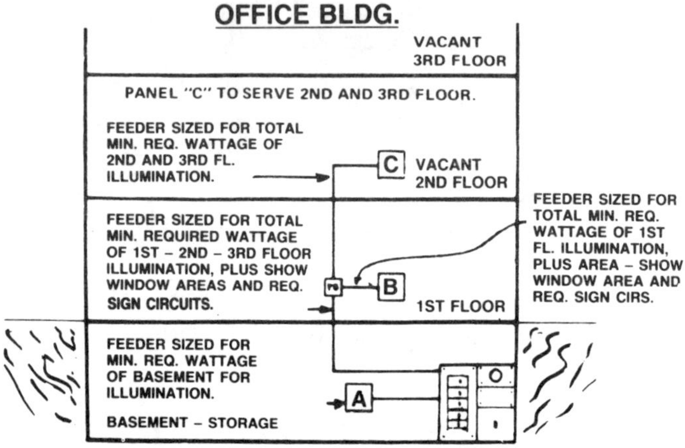

N.E.C. Sections 220-11 [& 220-3 (b)]

Computation Of Feeder Load

COMPUTING FEEDER LOAD FOR OFFICE BLDG.

MINIMUM VA.	VA.	TOTAL
18,000 SQ. FT. OFFICE AREA LIGHTING @ 3½ PER SQ. FT. 1990 NEC. 220-3 (b)	63,000	
3,000 SQ. FT. HALLS @ ½ W. PER SQ. FT.	1,500	
CONTINUOUS LOADS 125% X	64,500	80,625

NOTE: FOR GENERAL PURPOSE RECEPTACLE OUTLETS AT LEAST 1 VOLT-AMP PER SQ. FOOT SHALL BE PROVIDED WHEN THE ACTUAL QUANTITY IS UNKNOWN.

General Method – Section 220-11, 16, 17, 18, 19

Two Or More Dwelling Units Example

10 UNIT APT BLDG 1500 SQ FT EACH DWELLING

REGULAR

LIGHTING . . . . .4500x10 = 45000
APPLIANCE . . . . .3000x10 = 30000
75000

APPLICATION OF DEMAND FACTOR:

3000 @ 100% =	3000	
72,000 @ 35% =	25200	
	28200	28200 VA
D.W. 1500, G.D. 800 = 2300 x 10 = 23000 x 75% =		17250
A.C. 5000x10 =		50000
DRYER 5000x10 = 50000x50%		25000
E.R. 12KW (10) =		25000
* F.A.U. 825x10		*8250
		145450 VA / 240 V = 606 AMPS

*OMIT SMALLER OF DISSIMILAR LOADS

Optional (220-32)

LIGHT	45000 VA
SA	30000
APPLIANCES	23000
A/C	50000
DRYER	50000
*ELECTRIC RANGE	120,000
	318,000
	43% DEMAND
	136,740 / 140 V = 569.75 AMPS

*NOTE: THE SMALLER COMPUTED LOAD OF THE TWO METHODS MAY BE USED.

N.E.C. Section 220-13

Demand Factors For Receptacle Loads

RECEPTACLE DEMAND FOR NON-DWELLING INSTALLATIONS WHERE MORE THAN 10 KVA OF RECEPTACLES ARE INSTALLED, A DEMAND OF 50% IS PERMITTED FOR RECEPTACLES RATED AT 180 VA, FOR THAT PORTION OF THE LOAD OVER 10 KVA. EXAMPLES: OFFICE BUILDING HAVING 80 RECEPTACLE OUTLETS.

80 x 180 A =	14,400 VA
10 KW @ 100% =	10 KVA
4.4 KW @ 50% =	2.2 KVA
DEMAND LOAD =	12,200 VA

N.E.C. Section 220-19

Demand Load For Electric Range

COMPUTATIONS FOR ELECTRIC RANGE DEMAND 12KW RANGE — DEMAND LOAD = 8KW. FOR RANGE OVER 12KW INCREASE LOAD 5% FOR EACH ADDITIONAL KW OVER 12 & NOT OVER 27.

EXAMPLE

16.6 KW ELECTRIC RANGE
-12 KW
4.6 KW = 5 KW

12 KW DEMAND	= 8 KW
(5% OF 8 KW) × 5 (KW)	= 2.0 KW
DEMAND LOAD	= 10 KW

$$\frac{10{,}000}{240} = 41.66 \text{ AMPS}$$

N.E.C. Section 210-19 (b)

Minimum Rating For Fixed Electric Cooking Equipment

40 AMPERE CIRCUIT

(MINIMUM) FOR ELECTRIC COOKING APPLIANCES HAVING A RATING OF 8¾ KW OR LARGER

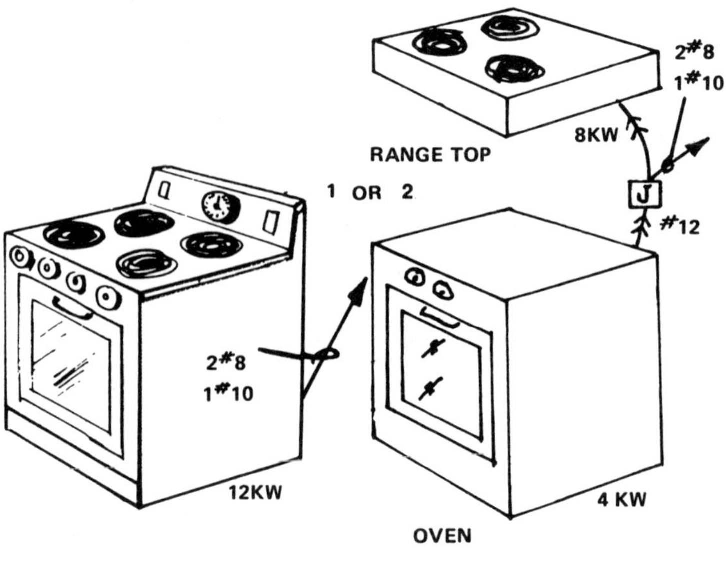

N.E.C. Sections 220-15, 220-21 & 220-30

Special Demands On Feeders Or Services

WHERE REFRIGERATED AIR CONDITIONERS AND SPACE HEATERS ARE LOCATED IN A SINGLE FAMILY DWELLING OR DWELLING UNIT — THE SMALLER OF THE TWO LOADS MAY BE OMITTED. DEMANDS ARE APPLIED BEFORE DETERMINING THE SMALLER LOAD.

UNIT A	
	3 SPACE HEATERS – TOTAL WATTS 4800W. OMIT
	1 REFRIGERATED AIR COND UNIT 5000 W.
	LOAD TO BE CALCULATED AT 100% 5000 W.

WHEN USING OPTIONAL METHOD FOR LOAD CALCULATION A 40% DEMAND SHALL BE APPLIED TO 4 OR MORE SEPARATELY CONTROLLED SPACE HEATERS BEFORE COMPARING LOADS.

UNIT B	
	5-1240 WATT SPACE HEATERS
	TOTAL 5x1240=6200 @ 40%=2480 W
	1 REFRIGERATED A/C UNIT=5000 W
	LOAD TO BE CALCULATED @ 100%=5000 W

N.E.C. Sections 220-21

Nonooincident Load

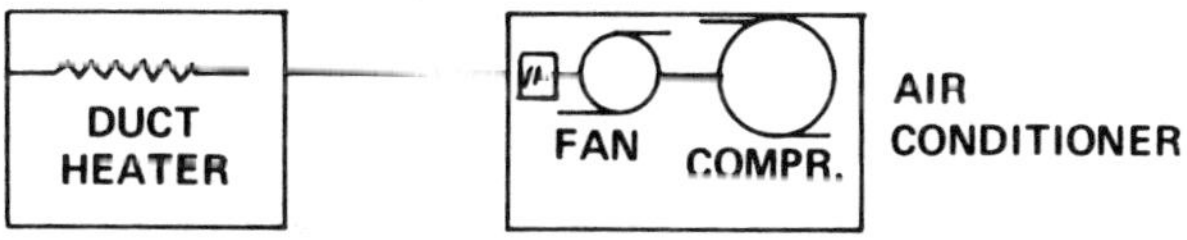

TO DETERMINE FEEDER LOAD, THE SMALLER OF THE 2 DISSIMILAR LOADS MAY BE OMITTED FROM THE TOTAL WHERE IT IS *UNLIKELY* THAT BOTH OF THE LOADS WILL BE SERVED SIMULTANEOUSLY.

N.E.C. Section 220-22

Neutral Conductor Size

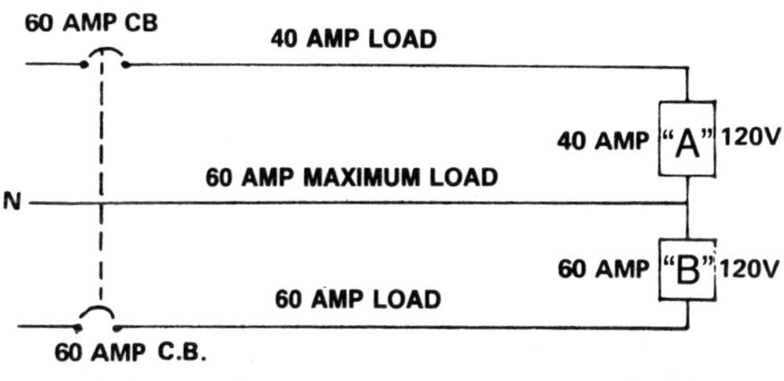

NEUTRAL CONDUCTORS SIZE IS 60 AMPS IF THE 40 AMP LOAD IS SHUT OFF. NOTE: "A" AND "B" REPRESENT TOTAL LINE TO NEUTRAL LOADS.

N.E.C. Section 220-22

Reduce Neutral Size

4 WIRE FEEDER

REDUCED NEUTRAL

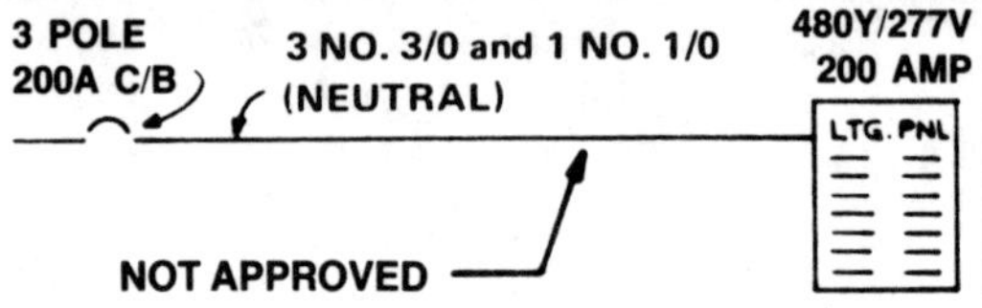

THE 1/0 NEUTRAL IS RATED FOR 150 AMPERES.

THE MAXIMUM UNBALANCE COULD BE 200 AMPERES (IF NO ADDITIONAL INFORMATION IS PROVIDED).

IF LOADS ARE KNOWN SEE EXAMPLE BELOW.

N.E.C. Section 220-22

Neutral Conductor Size

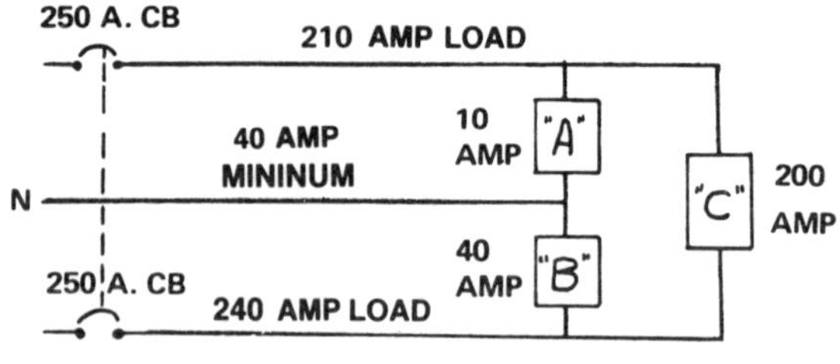

MAXIMUM UNBALANCE IS 40 AMPS.
MINIMUM NEUTRAL IN THIS CASE = 40 AMPS.
NOTE: "A", "B" AND "C" ARE SHOWN TO REPRESENT DISTRIBUTION OF LINE-TO-LINE AND LINE-TO-NEUTRAL PARTS OF THE TOTAL LOAD.

THERE SHALL BE NO REDUCTION OF THE NEUTRAL CAPACITY FOR THAT PORTION OF THE LOAD WHICH CONSISTS OF ELECTRIC-DISCHARGE LIGHTING, DATA PROCESSING OR SIMILAR EQUIPMENT, AND SUPPLIED FROM A 4-WIRE, WYE-CONNECTED 3-PHASE SYSTEM.

N.E.C. Section 220-30

Single Dwelling Feeder (Service) Optional Table Method 100 A-3 Wire

100 AMP — 3 WIRE OR 70 AMP 4 WIRE FEEDER

	LOAD	DEMAND FACTOR
SELECT LARGEST OF THESE FOUR	NAMEPLATE RATING OF AIR CONDITIONER AND AIR HANDLING EQUIPMENT	100%
	NAMEPLATE RATING OF CENTRAL ELECTRICAL SPACE HEATING	65%
	LESS THAN 4 ELECTRICAL SPACE HEATING UNITS	65%
	NAMEPLATE RATING OF ELECTRIC THERMAL STORAGE AND OTHER CONTINUOUS TYPE HEATING SYSTEMS	100%
	FOUR OR MORE SEPARATELY CONTROLLED SPACE HEATING UNITS	40%
	FIRST 10 KVA OF ALL OTHER LOADS	100%
	REMAINDER OF ALL OTHER LOADS	40%

ALL OTHER LOADS INCLUDES:

1 – BRANCH CIRCUIT LOADS LISTED IN SEC. 220-16

2 – THE NAMEPLATE RATING OF ALL APPLIANCES AND UTILIZATION EQUIPMENT SUPPLIED BY A SPECIFIC APPLIANCE BRANCH CIRCUIT NOT LISTED ABOVE.

(SEE SEC. 220-21 FOR LOADS THAT CAN BE OMITTED.)

10 Unit General (Without Ranges)

LIGHTS & PLUGS	28,200 VA
APPLIANCES	17,250 VA
A/C	50,000 VA
DRYER	25,000 VA
	120,450 VA

$$\frac{120{,}450 \text{ VA}}{240} = 501.8 \text{ AMPS}$$

N.E.C. Section 220-30

Computation Of Load For A Dwelling Unit 100 A-3 Wire Feeder Or Service Optional Method

CIRCUITS	VA NOTE	VA WITH DEMAND
1100 SQ. FT. AT 3VA PER SQ. FT. (2 LIGHT AND 1 CONVENIENCE OUTLET CIR).	3300	
2 SMALL APPLIANCES 1500 VA EACH	3000	
1 OUTDOOR	1500	
1 1/3 HP DISPOSAL	864	
1 BUILT-IN MIXER	380	
1 ELECTRIC DRYER	5000	
1 WATER HEATER	3000	
2 ATTIC FANS 750 VA EACH	1500	
1 3/4 HP POOL MOTOR AND 500 VA LIGHT	2156	*
1 10.25 KVA RANGE	10,250	
	30,950 VA	
FIRST 10 KW at 100%		10000
BALANCE OF LOAD (20950W) AT 40%		8380
3 SPACE HEATERS AT 65% (2.8)–1.2–1.2 KW)	OMIT	3380 ▲
1 CENTRAL AIR CONDITIONER		5000 ▲
		23,380 VA

▲ SEE SEC. 220-21
*SEE SEC 220-14

N.E.C. Section 220-32

Optional (10 Units)

LITES & PLUGS	45000
S.A.	30000
APP.	23000
A/C	50000
DRYER	50000
*(E.R.) 8 KWx10	80000
	278000W
	@43%

$$\frac{119540}{240} = 498 \text{ AMPS}$$

N.E.C. Section 225-6 (a)

Minimum Size Of Span Wires

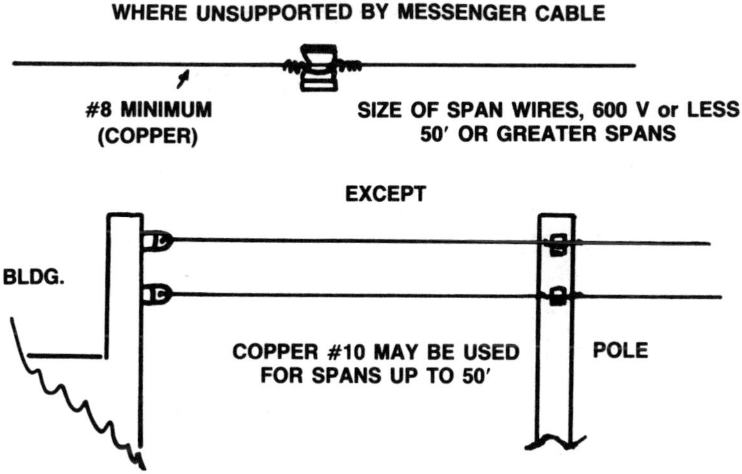

N.E.C. Sections 225-6 (b) &225-13

Festoons

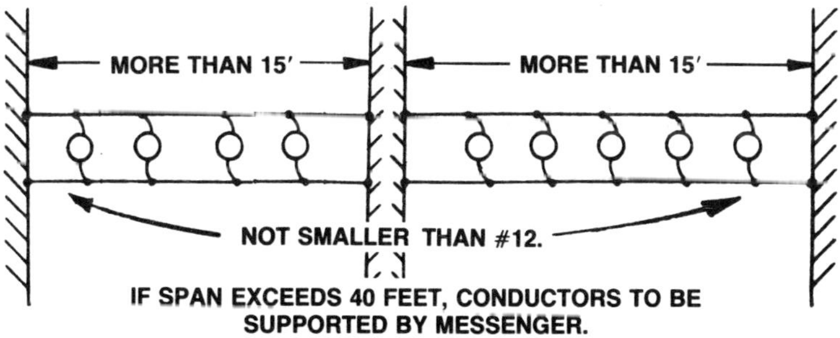

N.E.C. Sections 225-7 & 225-18

Outside Wiring

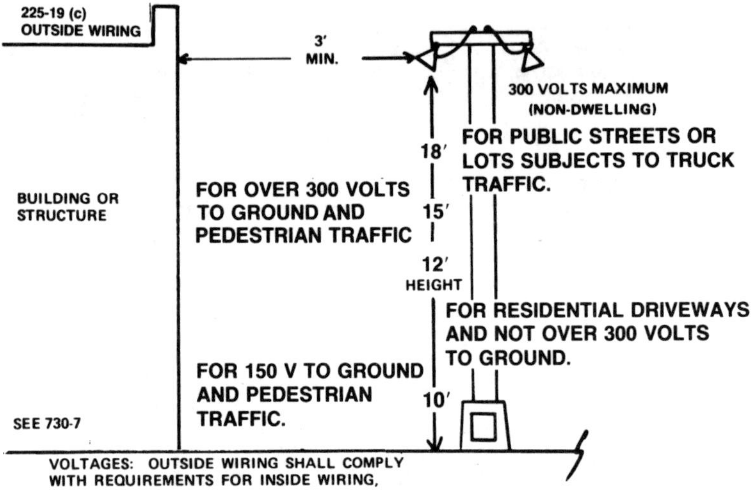

VOLTAGES: OUTSIDE WIRING SHALL COMPLY WITH REQUIREMENTS FOR INSIDE WIRING, EXCEPT: NO VOLTAGE LIMITATION WITH ABOVE HORIZONTAL AND VERTICAL CLEARANCES.

N.E.C. Section 225-9

Overcurrent Protection For Outside Wiring

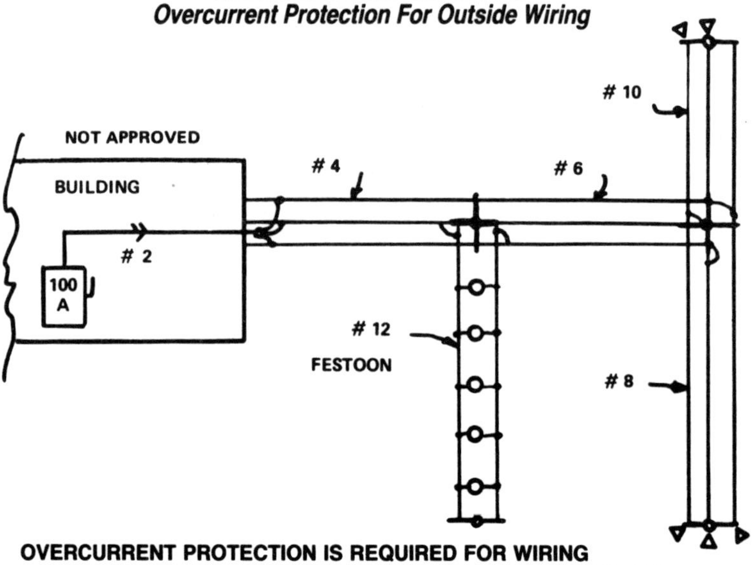

OVERCURRENT PROTECTION IS REQUIRED FOR WIRING SUPPLYING LIGHTING EQUIPMENT. SEE SEC. 210-20 & ART. 240.

N.E.C. Section 225-14

Permissible Use Of Open Wiring

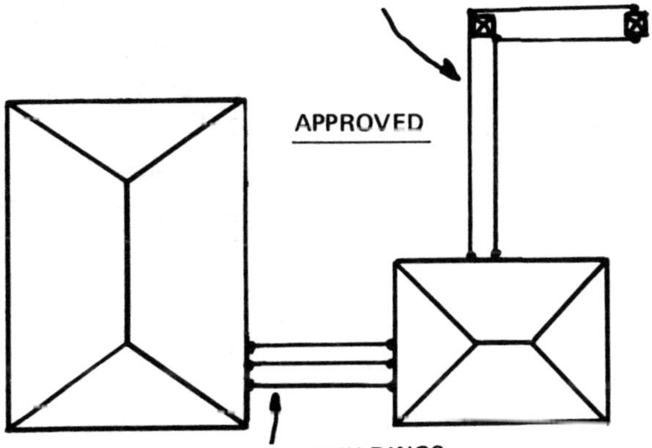

FOR ATTACHMENT
SEE SECTION 225-17

N.E.C. Section 225-14 (d)

Spacing Between Span Wires (600 V. Or Less)

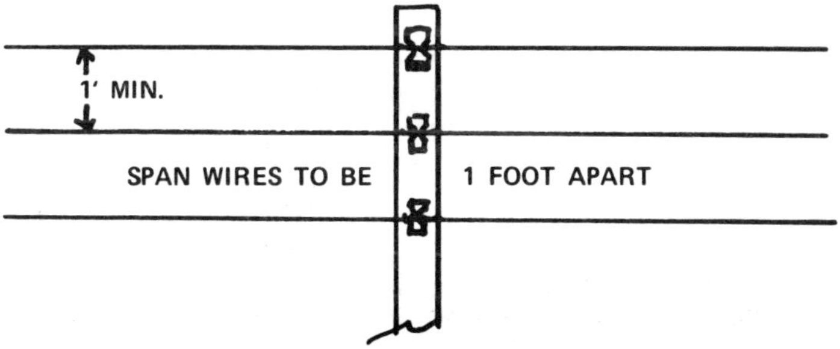

N.E.C. Section 225-19

Clearance Of Overhead Wires 0-300 Volts

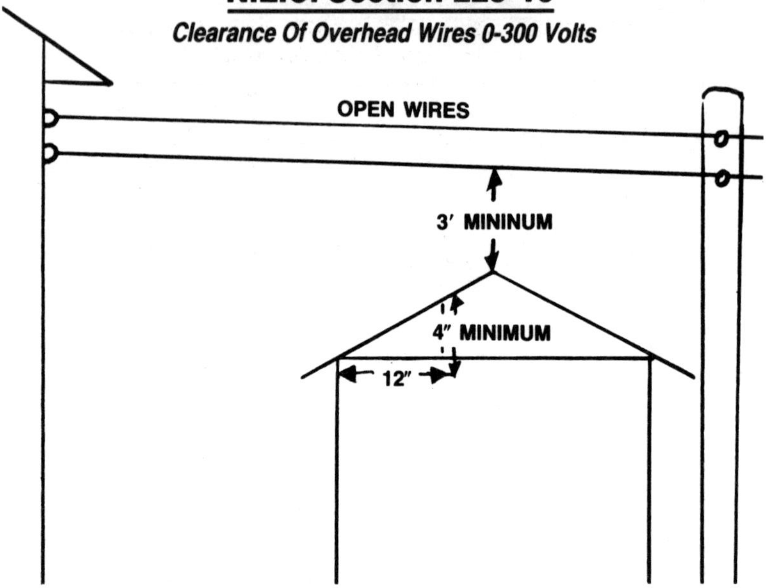

3′ CLEARANCE O.K. OVER BUILDING WITH RISE OF 4″ OR OVER PER FOOT.

N.E.C. Section 225-19 (a), Exc. #1

Clearance Of Open Wires From Sun Deck, Etc.
(For Insulated Conductors)

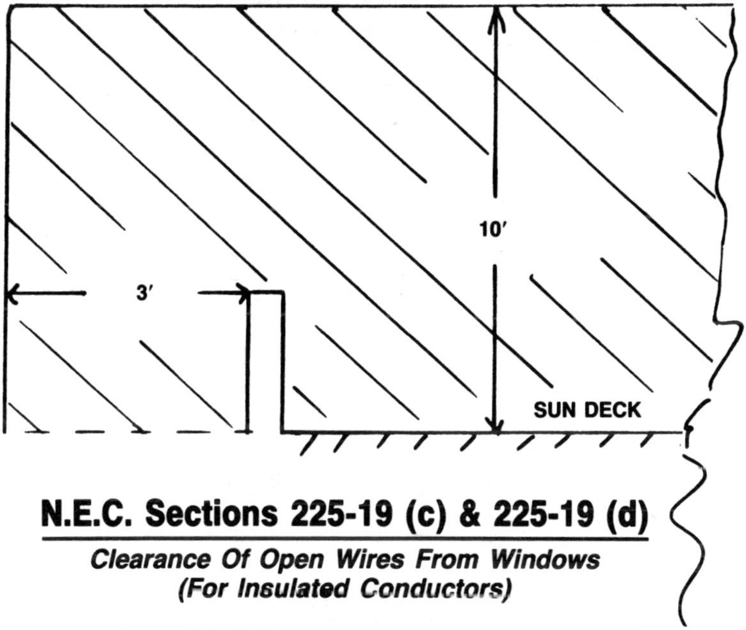

N.E.C. Sections 225-19 (c) & 225-19 (d)

Clearance Of Open Wires From Windows
(For Insulated Conductors)

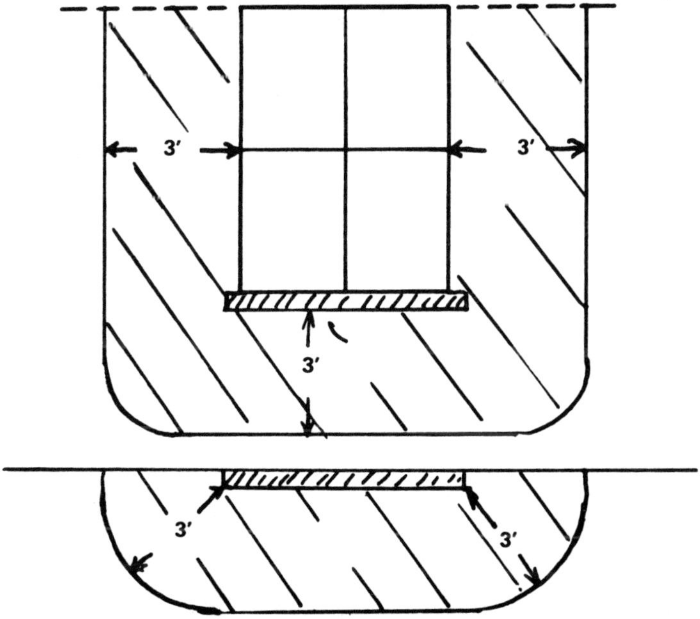

N.E.C. Section 225-25

Location Of Lamps

LAMPS FOR OUTDOOR LIGHTING SHALL BE LOCATED BELOW ALL LIVE CONDUCTORS, TRANSFORMERS OR OTHER ELECTRICAL EQUIPMENT.

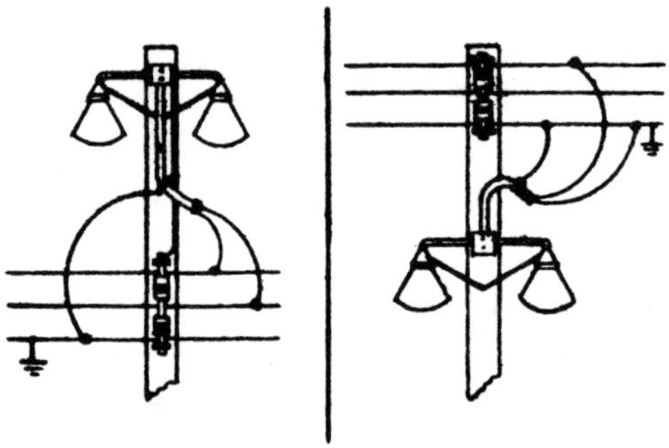

NOT APPROVED APPROVED

N.E.C. Section 225-26

Live Vegetation

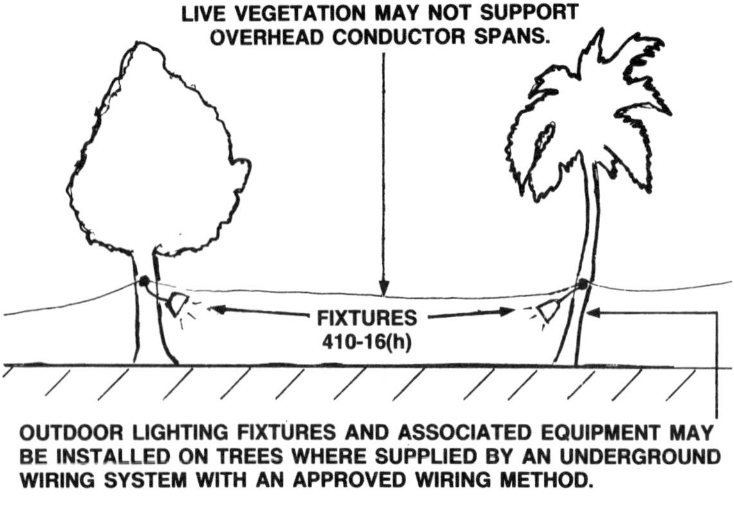

N.E.C. Section 230-2 & 230-40, Ex. No. 1 & No. 2

Service Entrance Conductors And Subsets Of Service Entrance Conductors For Multiple Occupancy Buildings

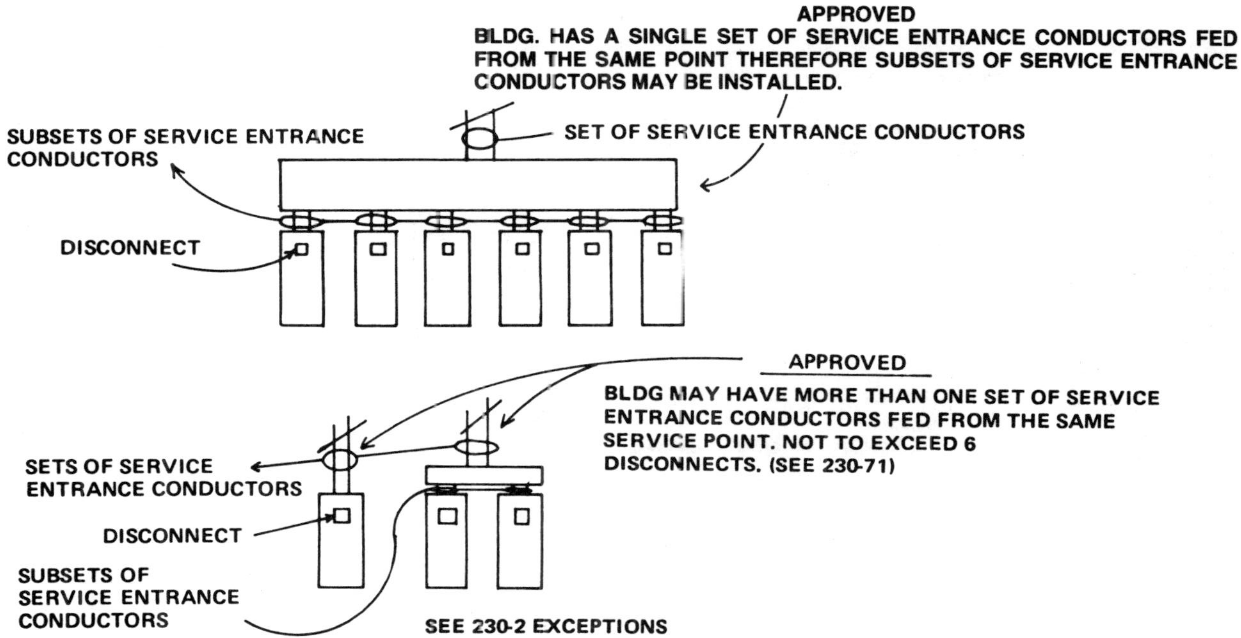

N.E.C. Sections 230-2 & 230-40, Ex. 1 & 2

Service Entrance Conductors For Multiple Occupancy Buildings

NOTE: THIS DOES NOT MODIFY DISCONNECT REQUIREMENTS.

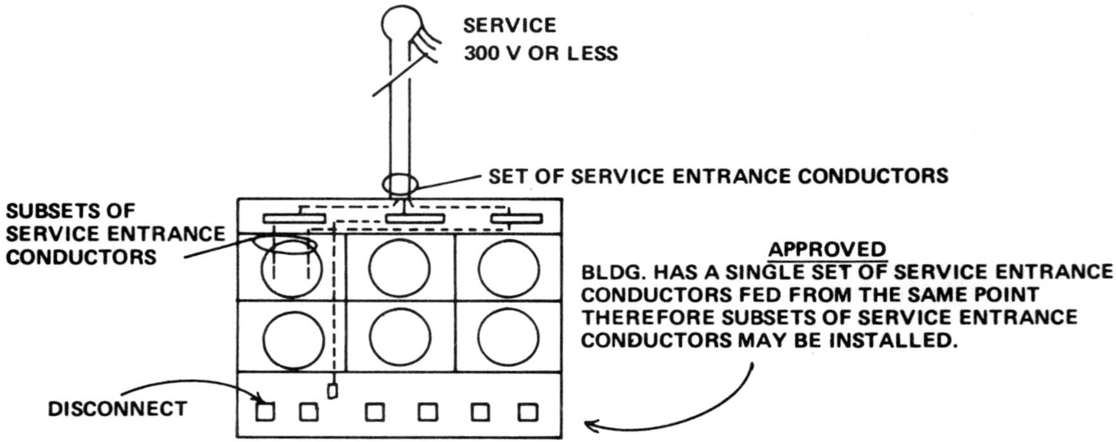

N.E.C. Section 230-2, Exception 4 (b)

Number of Services

Single Phase Only

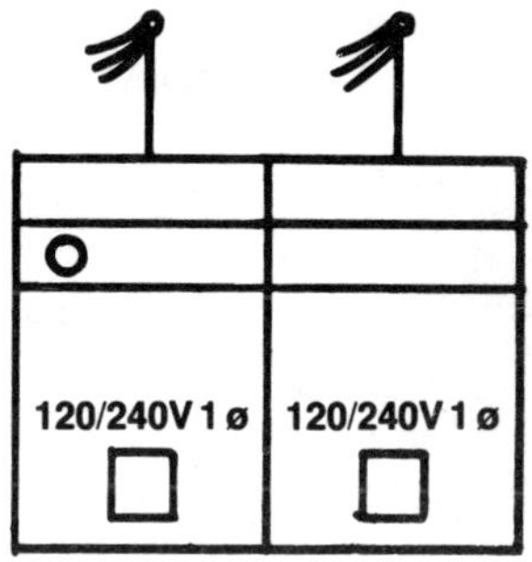

ONE CONSUMER MAY HAVE MORE THAN ONE SERVICE FOR EACH CLASS OF SERVICE WHERE THE TOTAL LOAD EXCEEDS SERVING AGENCY REQUIREMENTS.

N.E.C. Section 230-2, Exception 4 (a)

Capacity Requirements

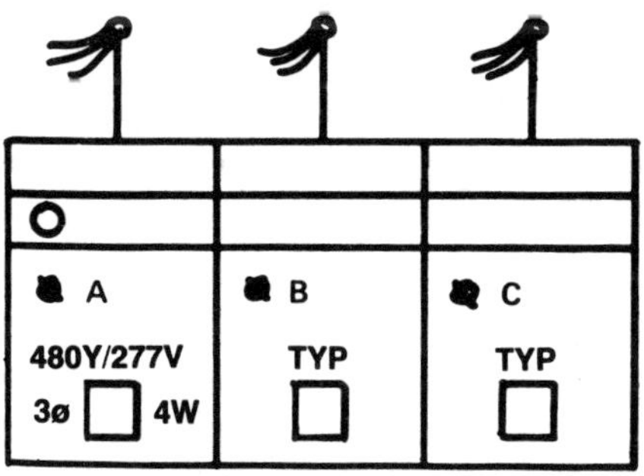

ONE CONSUMER MAY HAVE TWO OR MORE SETS FOR EACH CLASS OF SERVICE WHERE THE TOTAL LOAD EXCEEDS 2000 AMPERES.

METERING EQUIPMENT:
THE SERVICING AGENCY IS TO HAVE JURISDICTION OVER THE SIZE, LOCATION AND TYPE OF METER SOCKETS AND CURRENT TRANSFORMERS BOXES.

N.E.C. Section 230-2, Exceptions 5 & 6

Number Of Services

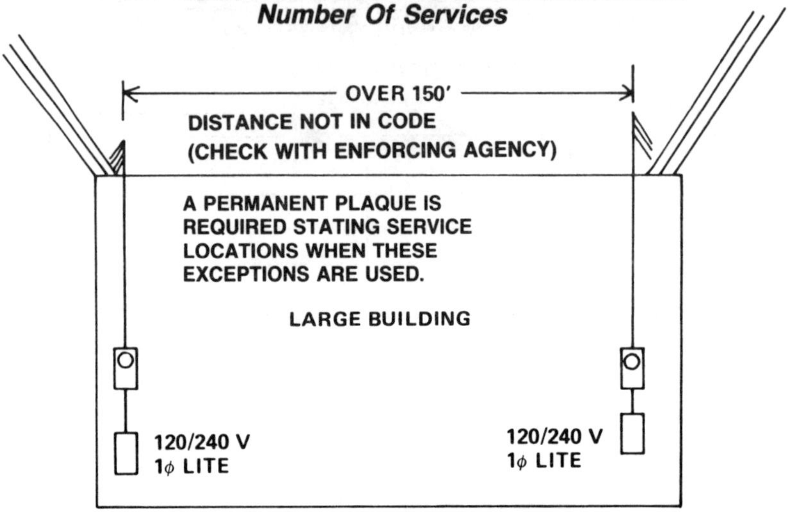

MEET THE SERVING AGENCIES REQUIREMENTS AND PROVIDE DIFFERENT CHARACTERISTICS OR CLASS OF USE.

NOTE: THE SERVING AGENCY SHOULD BE CONSULTED REGARDING ITS REGULATIONS.

N.E.C. Section 230-2, Exception 6

(More Than One Service)

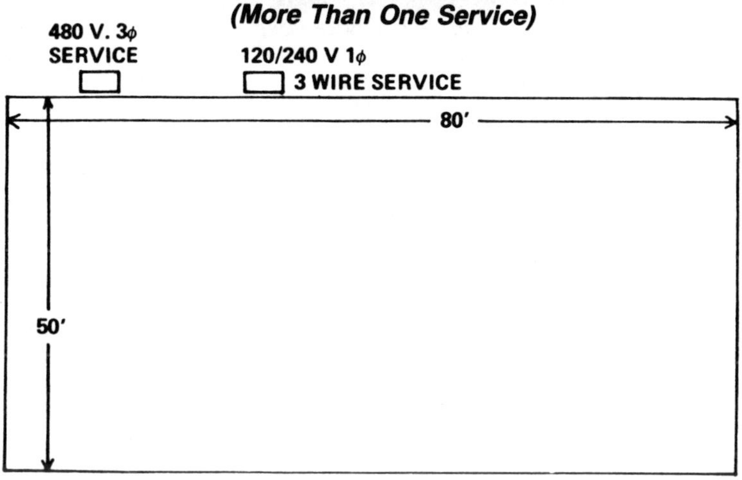

ADDITIONAL SERVICE POINTS MAY BE ESTABLISHED FOR DIFFERENT CHARACTERISTICS OR CLASSES OF USE.

N.E.C. Sections 230-2, Ex. 6

Number Of Services Different Classes

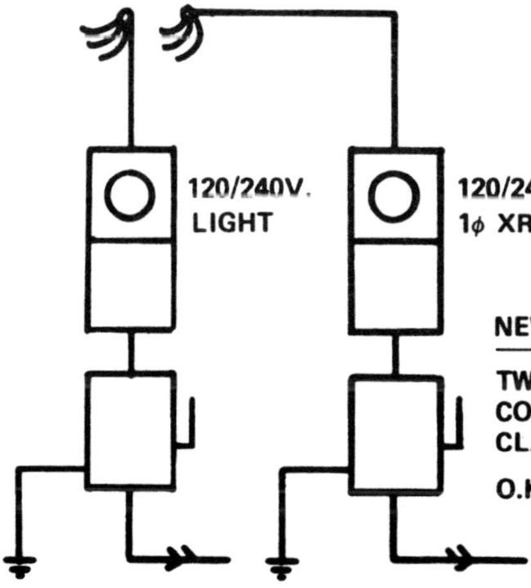

NEW OR OLD

TWO SETS OF SERVICE ENTRANCE CONDUCTORS FOR DIFFERENT CLASSES OF LOAD.

O.K.

N.E.C. Sections 230-2, Ex., 230-40

More Than One Set Of Service Entrance Conductors From One Service Point

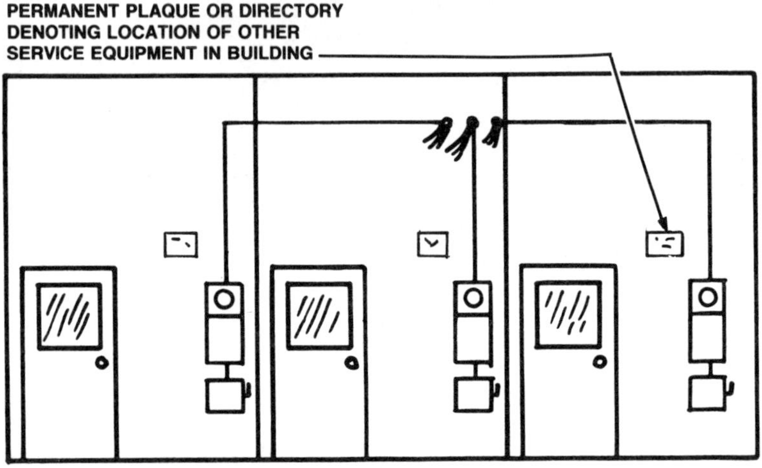

MULTIPLE OCCUPANCIES

SERVICE ENTRANCE CONDUCTORS MAY BE SUPPLIED FROM ONE POINT OF ATTACHMENT FOR MULTIPLE OCCUPANCIES. THE SERVING AGENCY MAY PROHIBIT MORE THAN 3 SETS OF SERVICE ENTRANCE CONDUCTORS AT THE POINT. A PERMANENT PLAQUE IS REQUIRED STATING SERVICE LOCATIONS.

N.E.C. Section 230-42

Dwellings — Minimum Service-Entrance Conductor Sizes

100 AMPERE SUPPLY

FOR – (MINIMUM SIZE 3 WIRE SERVICE)

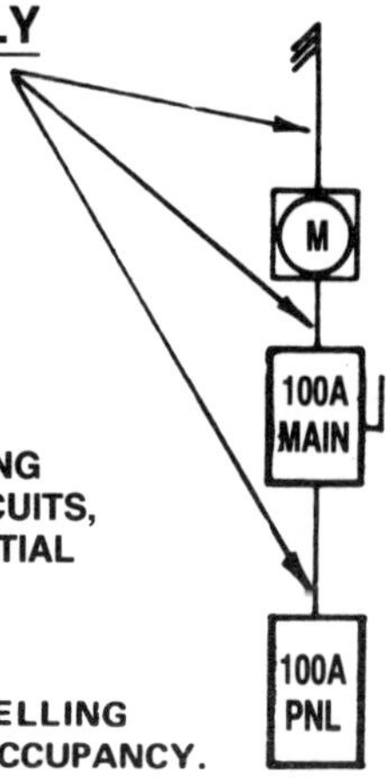

SINGLE FAMILY DWELLING WITH SIX OR MORE CIRCUITS, OR 10 KVA OR MORE INITIAL COMPUTED LOAD

NOTE: THIS DOES NOT APPLY TO DWELLING UNITS IN BUILDINGS OF MULTIPLE OCCUPANCY.

N.E.C. Section 230-42

Service Entrance Conductors Single Family

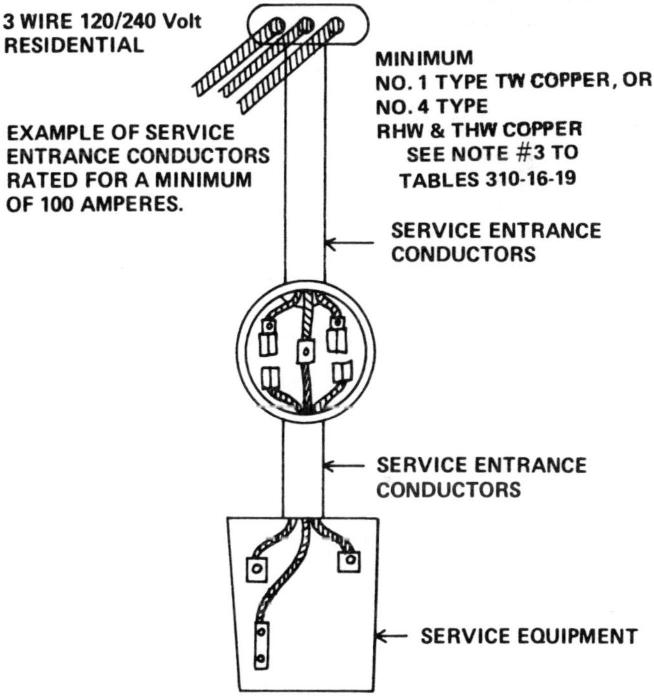

N.E.C. Section 230-43

Wiring Methods

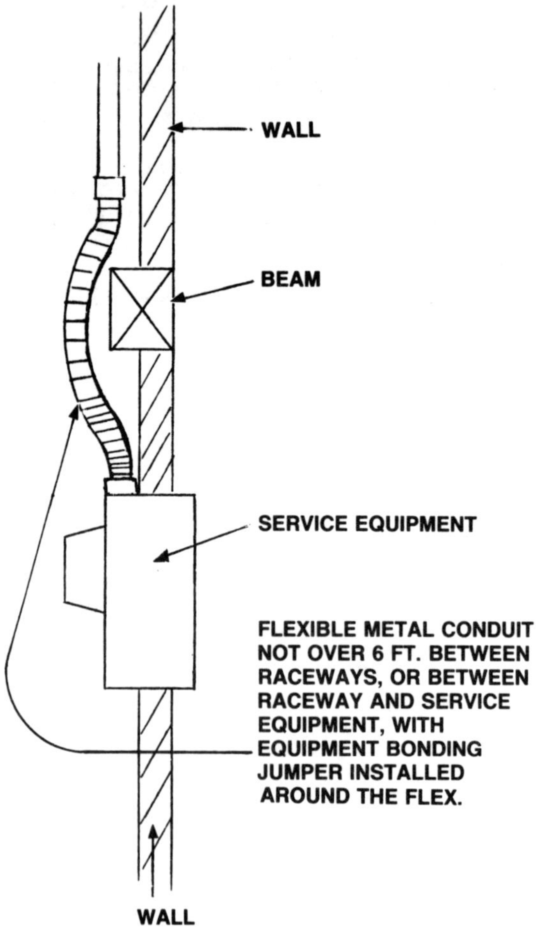

N.E.C. Section 230-53

Raceways To Drain

WHERE EXPOSED TO THE WEATHER, OR EMBEDDED IN MASONRY RACEWAYS ENCLOSING SERVICE ENTRANCE CONDUCTORS SHALL BE RAINTIGHT AND ARRANGED TO DRAIN.

FOR SERVICE HEAD
SEE: 230-54

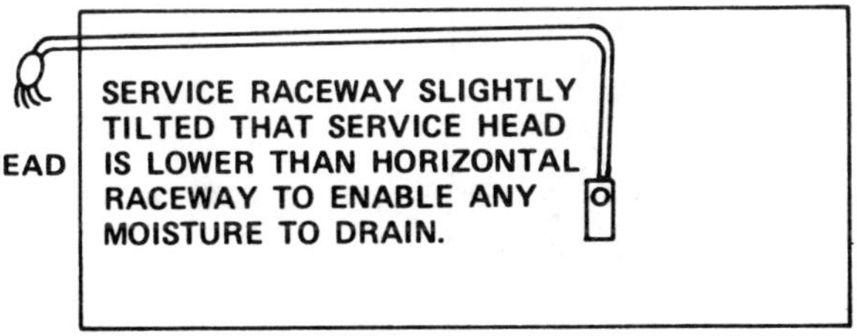

N.E.C. Sections 230-71, 384-16 (a), Ex. 2

Exceptions To Requirements For Main Service Switch

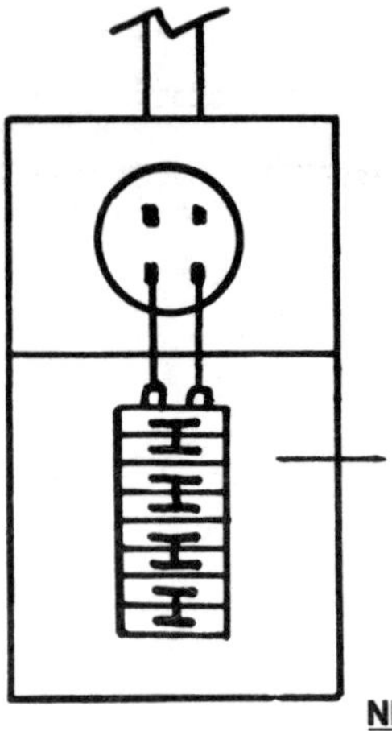

MAIN SWITCH NOT REQUIRED WHERE 6 OR LESS SWITCHES OR CIRCUIT BREAKERS ARE INSTALLED IN AN EXISTING PANELBOARD. FOR THE PURPOSE OF THIS SECTION A HANDLE TIE IS CONSIDERED AS 1 CIRCUIT. BUT . . .

NEW LIGHTING AND APPLIANCE SERVICE PANELBOARDS MUST BE PROTECTED BY NOT MORE THAN TWO MAIN BREAKERS OR TWO SET OF FUSES.

Service Disconnecting Means.

THE DISCONNECTING MEANS FOR A BUILDING MAY CONSIST OF NOT MORE THAN 6 SWITCHES OR CIRCUIT BREAKERS.

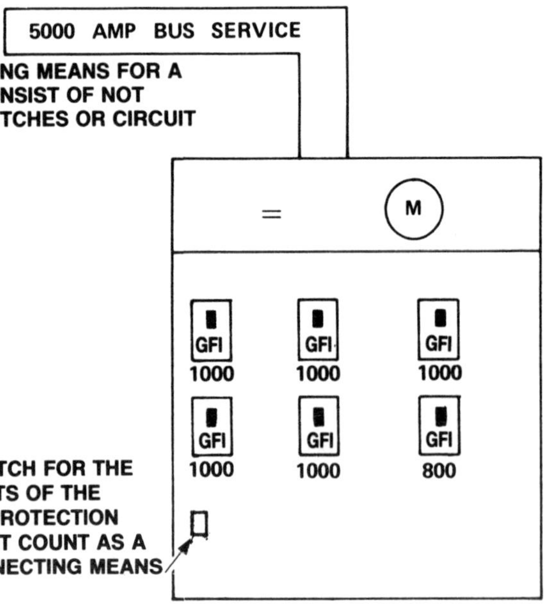

DISCONNECT SWITCH FOR THE CONTROL CIRCUITS OF THE GROUND-FAULT PROTECTION SYSTEM DOES NOT COUNT AS A SERVICE DISCONNECTING MEANS

SINGLE MAIN DISCONNECT NOT REQUIRED.

N.E.C. Section 230-71

Main Switch Required

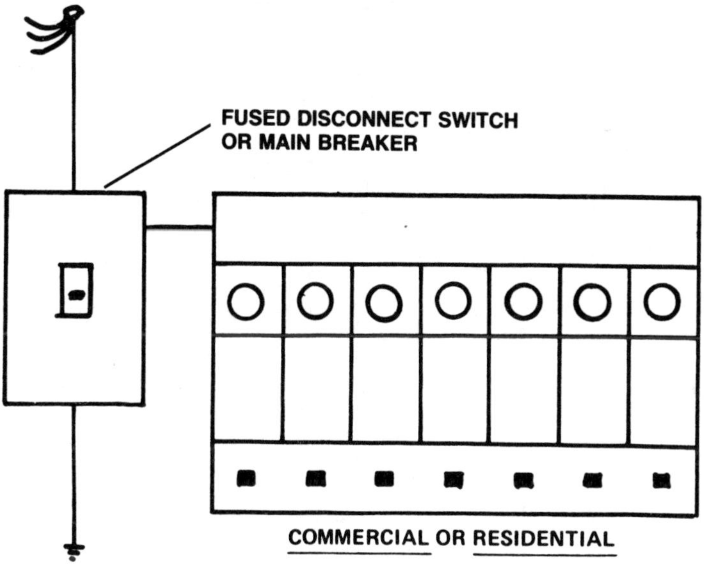

COMMERCIAL OR RESIDENTIAL

MORE THAN *SIX* SETS OF SERVICE EQUIPMENT

N.E.C. Section 230-71 (a)

Main Switch Not Required

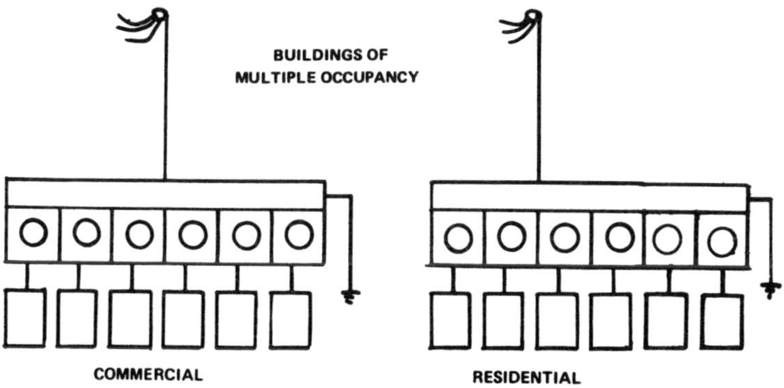

SINGLE MAIN SWITCH NOT REQUIRED. NOT MORE THAN SIX SETS OF SERVICE EQUIPMENT WITH MAINS WHERE GROUPED.

N.E.C. Section 230-71 (a)

Article 100 — Definition of "Service"

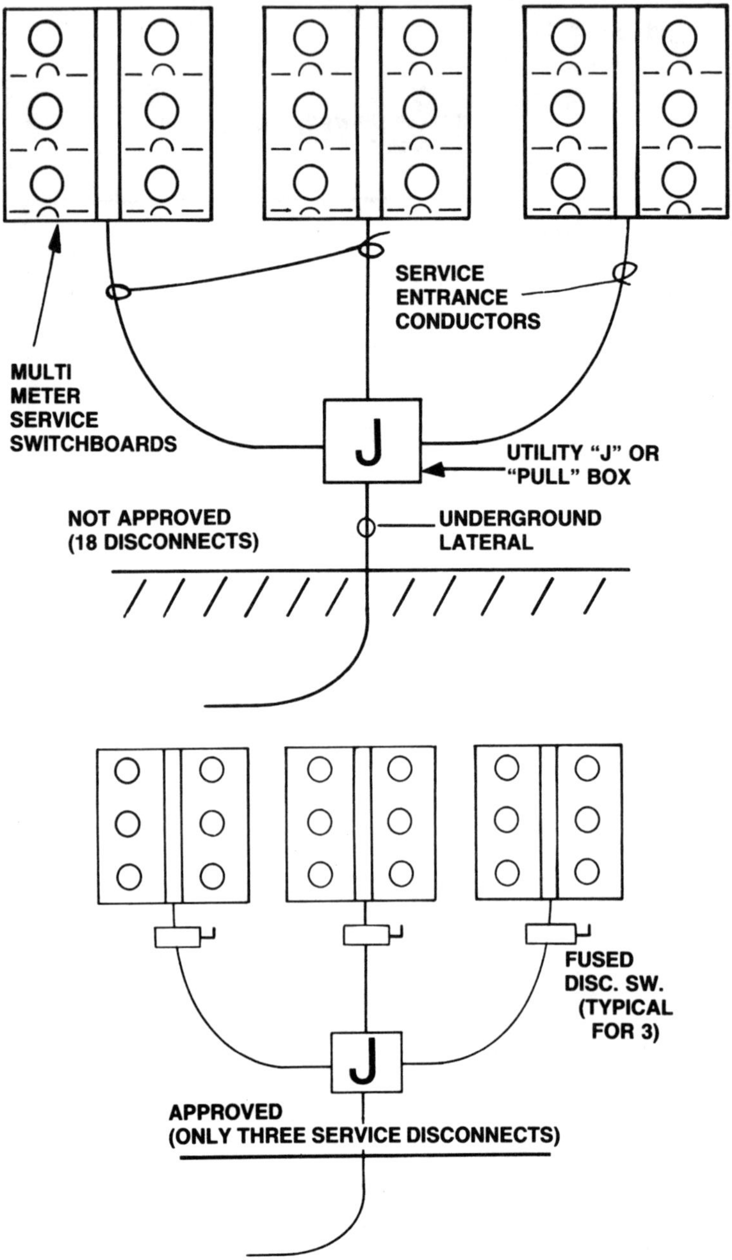

N.E.C. Section 230-72 (b)

Service Equipment Grouped

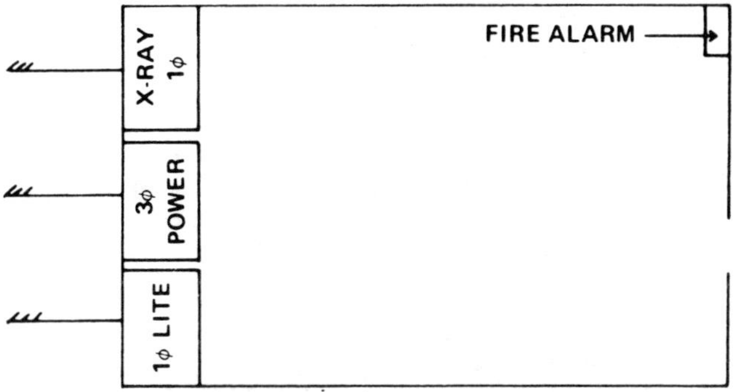

WHERE SUPPLIED BY THE SAME SIDE OF THE BUILDING BY MORE THAN ONE SET OF SERVICE ENTRANCE CONDUCTORS THE EQUIPMENT SHALL BE GROUPED AND MARKED TO INDICATE LOAD SERVED.

THE ONE OR MORE DISCONNECTING MEANS FOR FIRE PUMPS OR FOR EMERGENCY SERVICES SHALL BE INSTALLED SUFFICIENTLY REMOTE FROM THE SERVICE DISCONNECTING MEANS FOR THE NORMAL SERVICES TO MINIMIZE POSSIBILITY OF SIMULTANEOUS INTERRUPTION OF SUPPLY.

N.E.C. Section 230-70 (a)

Service Switch Location

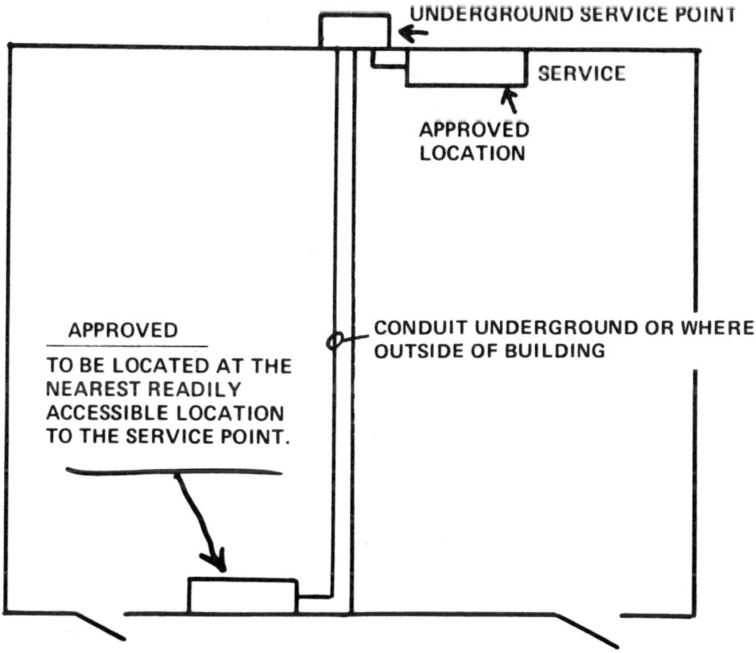

N.E.C. Section 230-79

Minimum Service Switch Size Other Than Specified In (a), (b) Or (c)

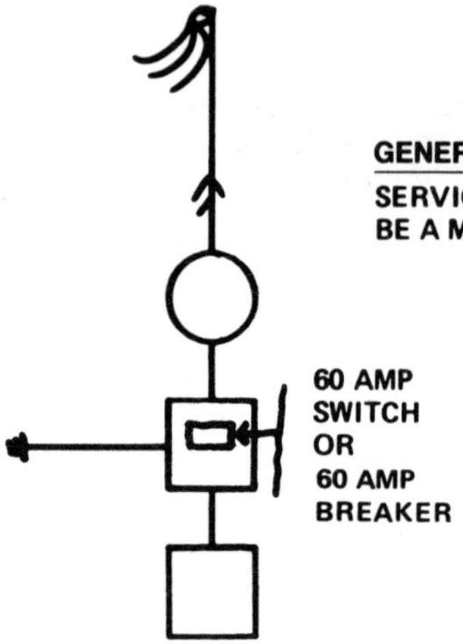

GENERAL REQUIREMENT [230-79 (d)]

SERVICE DISCONNECTING MEANS TO BE A MINIMUM OF 60 AMPS

N.E.C. Section 230-79 (a)

15 Amp. Minimum To One Circuit Of Limited Load

NOTE: 12 COPPER CONDUCTOR MINIMUM
SEE SEC. 230-42, 23(b), 230-31(b).

N.E.C. Section 230-79 (b)

Limited Load

EXCEPT

30 AMP SWITCH OR BREAKER APPROVED FOR:

1. NOT MORE THAN TWO- 2 WIRE CIRCUITS; OR
2. NOT MORE THAN ONE 3 WIRE CIRCUIT; OR
3. ONE POLYPHASE BRANCH CIRCUITS; OR
4. LOAD MAY BE ADDED TO SUITABLE EXISTING 30 AMP SWITCHES.

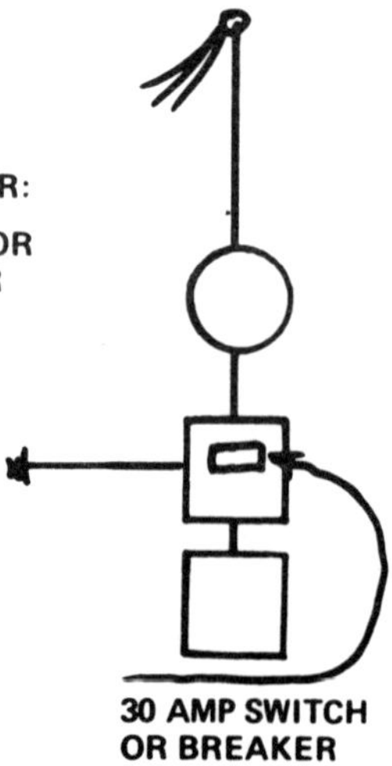

N.E.C. Section 230-79 (c)

Rating Of Service Disconnecting Means In Dwelling Panelboard

100 AMP. 3 WIRE PANELBOARDS FOR SINGLE FAMILY DWELLINGS OVER 10 KVA INITIAL LOAD OR OVER 5 INITIAL CIRCUITS.

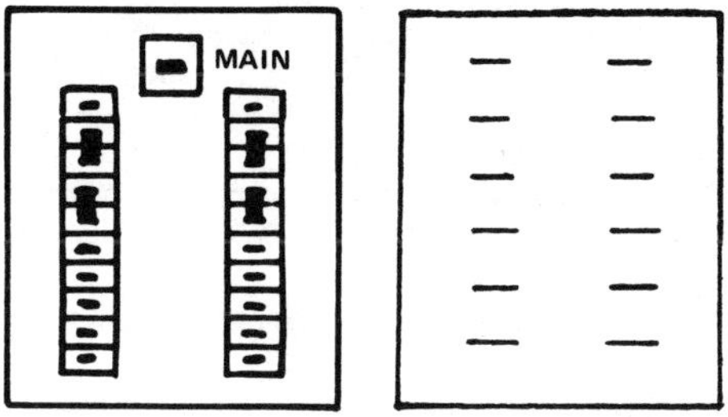

APPROVED

NO MAIN CIRC. BKR.
NOT APPROVED

N.E.C. Section 230-90

Overcurrent Protection

APPROVED

NOTE: FUSES OR BREAKER TO BE OF SIZE TO PROTECT SERVICE ENTRANCE CONDUCTORS AND FEEDER.

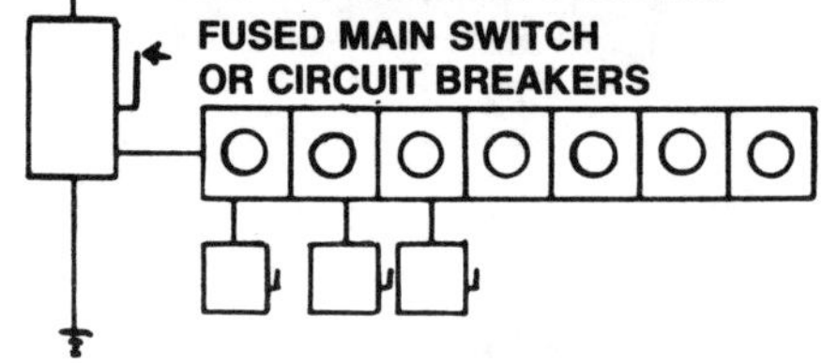

NOTE: SWITCH, FUSE, OR BREAKER MAY BE SIZED LARGER IF REQUIRED FOR MOTOR STARTING PROTECTION.

N.E.C. Section 230-94, Ex. 6

Electrically Operated Service Disconnects

WHERE SERVICE EQUIPMENT IS OPERATED ELECTRICALLY THE CONTROL CIRCUIT MAY BE CONNECTED AHEAD OF THE SERVICE EQUIPMENT IF SUITABLE OVERCURRENT PROTECTION AND DISCONNECTING MEANS ARE PROVIDED.

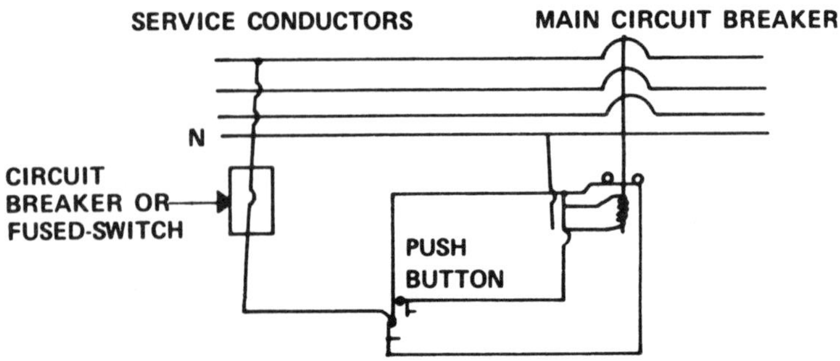

N.E.C. Section 230-95

Ground-Fault Protection Of Equipment

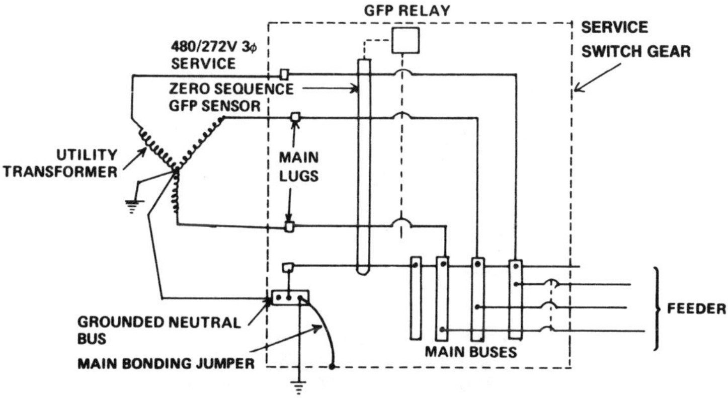

N.E.C. Section 230-95

Ground-Fault Protection Of Equipment

GROUND-FAULT PROTECTION OF EQUIPMENT SHALL BE PROVIDED FOR GROUNDED WYE ELECTRICAL SERVICES OF MORE THAN 150 VOLTS TO GROUND, SUCH AS 480/277 V 3Ø 4 WIRE SYSTEMS, BUT NOT EXCEEDING 600 VOLTS PHASE-TO-PHASE FOR ANY SERVICE DISCONNECTING MEANS RATED 1000 AMPERES OR MORE.

THE GROUND FAULT PROTECTION SYSTEM SHALL BE PERFORMANCE TESTED WHEN FIRST ENERGIZED WITH A WRITTEN RECORD TO BE SUBMITTED TO THE INSPECTION AGENCY.

THE GROUND FAULT PROTECTIVE DEVICE SHALL OPERATE TO CAUSE THE SERVICE DISCONNECTING MEANS TO OPEN ALL UNDERGROUND CONDUCTORS OF THE FAULTED CIRCUIT AT FAULT CURRENT VALUES OF 1200 AMPERES OR LESS, (MAXIMUM SETTING OF 1,200A).

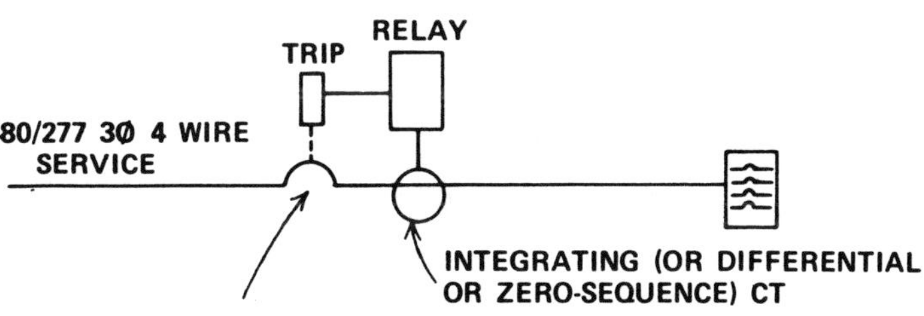

MAIN SERVICE CIRCUIT BREAKER WITH SHUNT TRIP, ELECTRICALLY OPERATED SWITCH OR BOLTED PRESSURE SWITCH WITH SHUNT TRIP.

N.E.C. Section 240-6

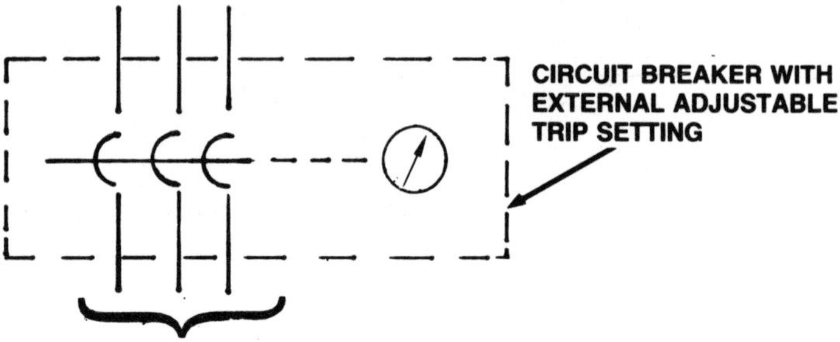

BRANCH CIRCUIT OR FEEDER CONDUCTOR MUST HAVE AN AMPACITY NOT LESS THAN THE MAXIMUM CIRCUIT BREAKER SETTING (SEE EXCEPTION ABOUT ADJUSTABLE SETTING) BEHIND LOCKED COVER.

N.E.C. Section 240-9

Thermal Relays Not Approved For Branch Circuit Protection

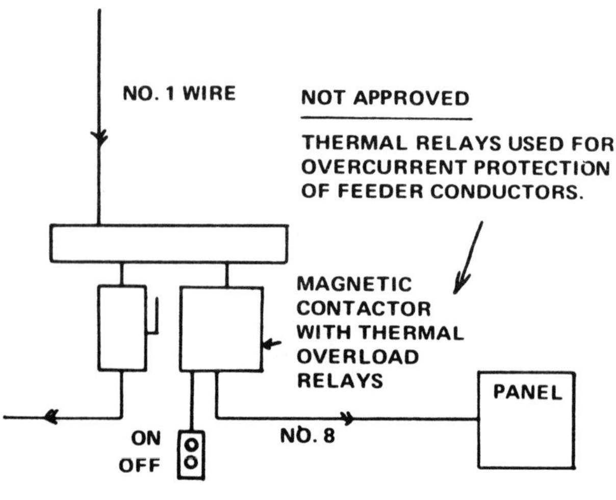

N.E.C. Section 240-21 (b)

10 Foot Tap Rule

FEEDER TAP SHALL BE NOT LESS THAN THE AMPERE RATING OF THE SWITCHBOARD OR PANELBOARD SUPPLIED BY THE TAP CONDUCTOR.

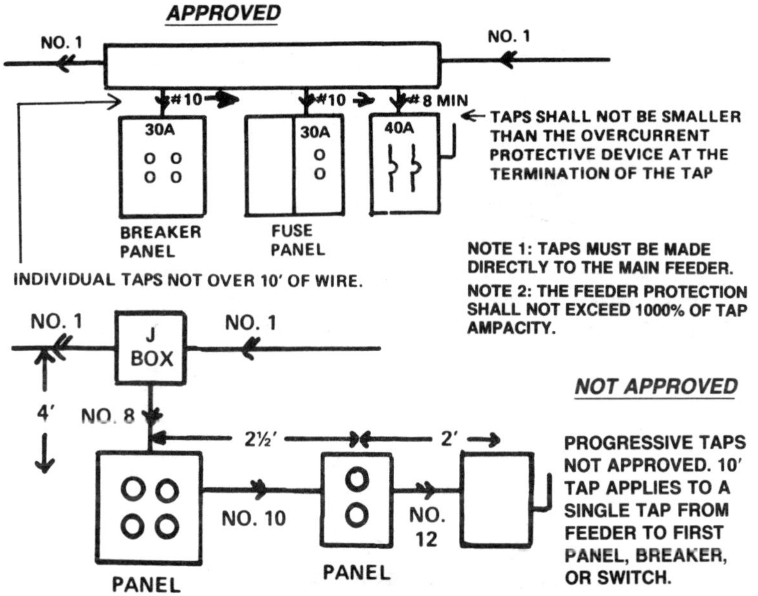

N.E.C. Section 240-21 (c)

25 Foot Tap Rule

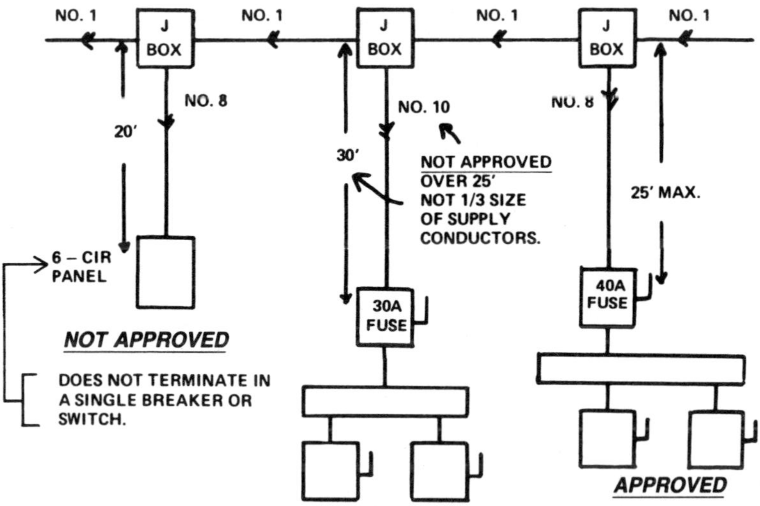

N.E.C. Section 240-21 (d) & 450-3 (b) 2

Transformer & Conductor Protection

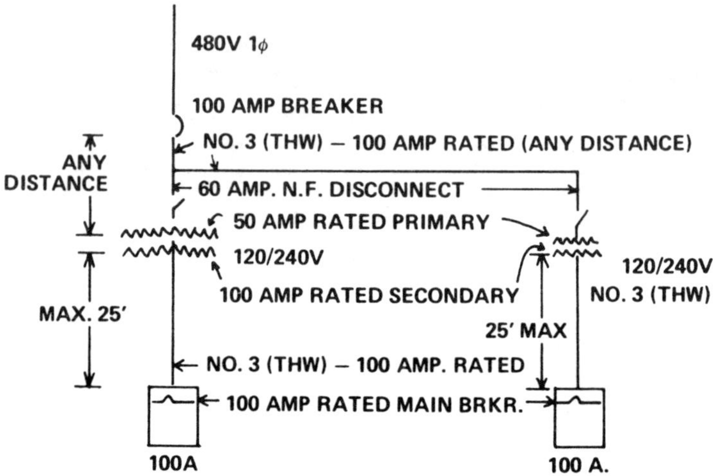

1. PRIMARY IS PROTECTED AT NOT TO EXCEED 250% OF PRIMARY RATING OF SMALLEST TRANSFORMER.
2. THE OVERCURRENT PROTECTION ON THE PRIMARY PROTECTS THE PRIMARY CONDUCTORS AT ITS RATED AMPACITY WHICH EXCLUDES THE 25' TAP RULE TO THE TRANSF. SECONDARY.
3. THE 100 AMP. RATED SECONDARY CONDUCTORS (NOT OVER 25' MAY BE 1/3 THE RATING OF THE OVERCURRENT PROTECTIVE DEVICE OR FEEDER AMPACITY FROM WHICH IT IS TAPPED. (2:1 RATIO 100 X 2=200A1/3 OF 200=66.6 AMPERES – NO. 3 THW IS LARGER THAN 1/3 TAP.)
4. TRANSFORMER SECONDARY CONDUCTORS, OTHER THAN 2 WIRE, ARE NOT CONSIDERED TO BE PROTECTED BY THE PRIMARY.

   100 AMP RATED OVERCURRENT DEVICE AHEAD OF EACH PANELBOARD IS NOT OVER 125% OF EACH TRANSFORMER SECONDARY RATING & LIMITS THE LOAD TO THE AMPACITY OF THE TAP.

240-6 STANDARD RATING OF FUSES 1, 3, 6, 10 AND 601 AMPERE RATED FUSES ARE CONSIDERED AS A STANDARD RATING.

N.E.C. Section 240-21 (e), 430-28, Ex.

**100 Foot Tap Rule*

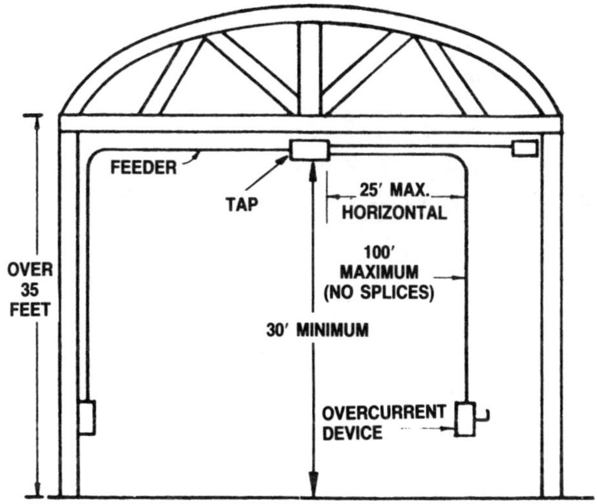

N.E.C. Section 240-21 (j), Ex. 11

25 Ft. Tap For Industrial Electrical System

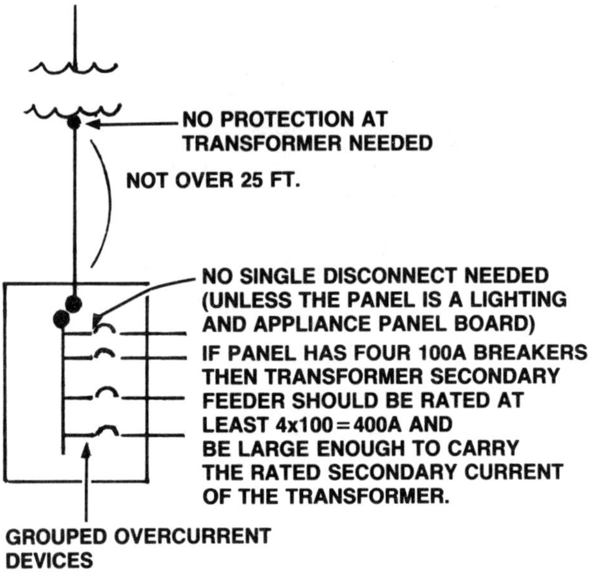

IF FEEDER IS 4-500 MCM THW AND THE TRANSFORMER IS 150 KVA, 480/120-208 V-3 PH. 4W. THE INSTALLATION WILL NOT COMPLY.

(THE SECONDARY RATED CURRENT IS 415A. THE AMPACITY OF 500 MCM, THW IS ONLY 380A.)

N.E.C. Section 240-24 (b)

Access To Panels

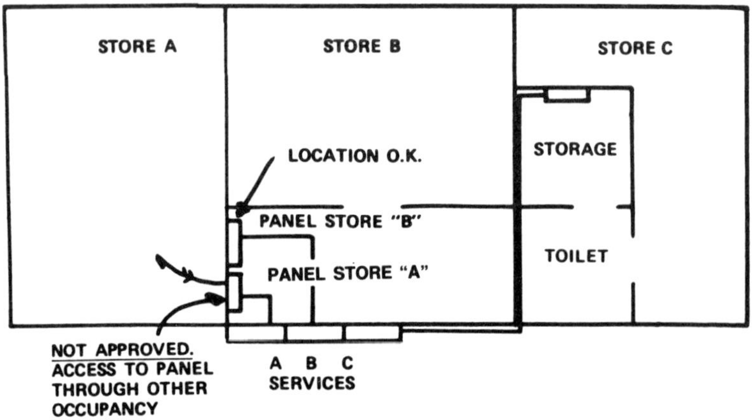

N.E.C. Section 240-40

Disconnect For Cartridge Fuses

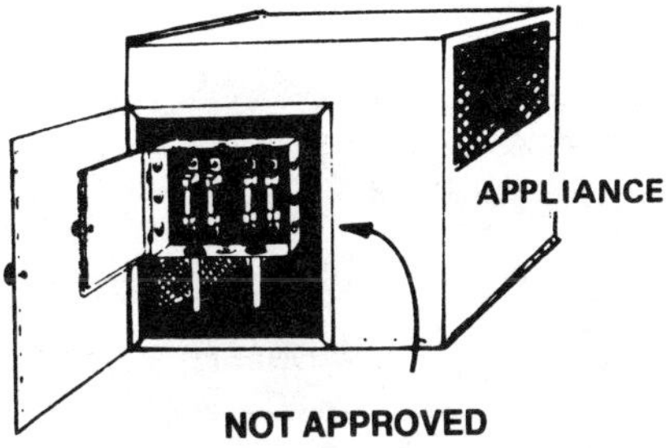

NO DISCONNECTING MEANS FOR CARTRIDGE FUSES . . .

N.E.C. Section 240-53

Type Of Plug Fuses Required

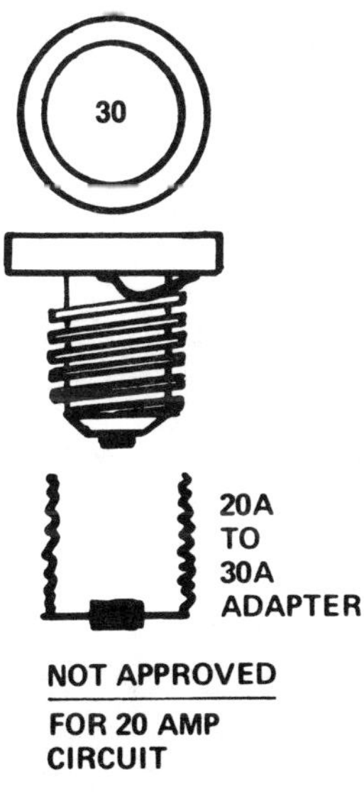

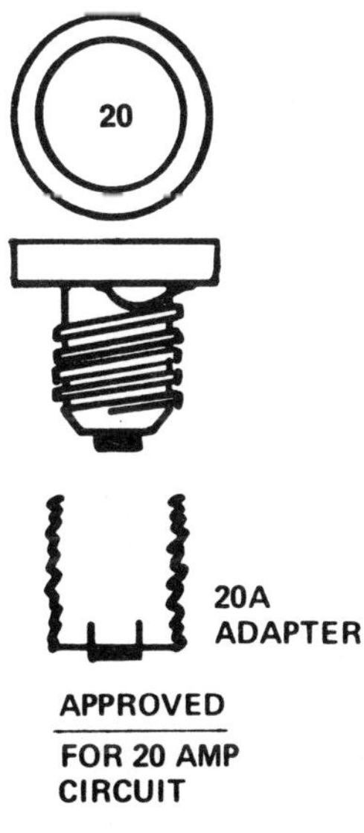

N.E.C. Section 240-53 & 54

Edison Base Fuses

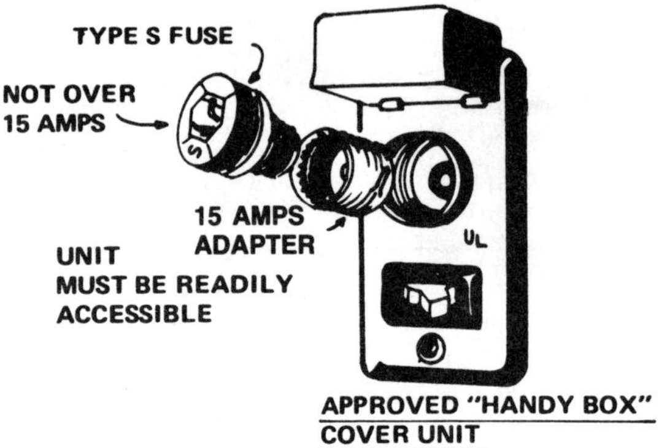

N.E.C. Section 240-83 (d)

Circuit Breaker Used As Switches

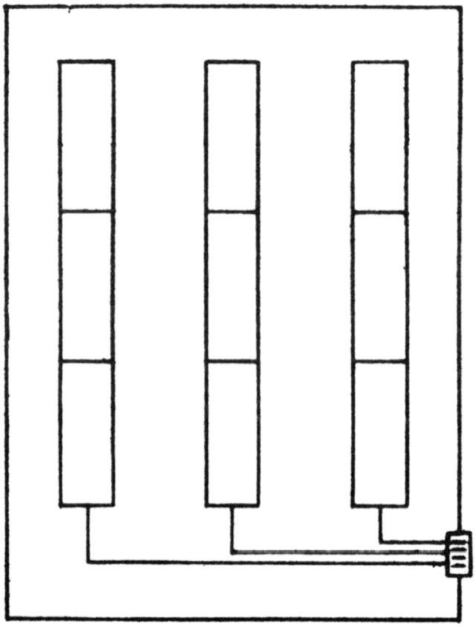

N.E.C. Sections 250-5(b)4 (Ex. 5) & 250-27

High-Impedance Grounded Neutral System Connections For A/C Systems 50V To 1,000V

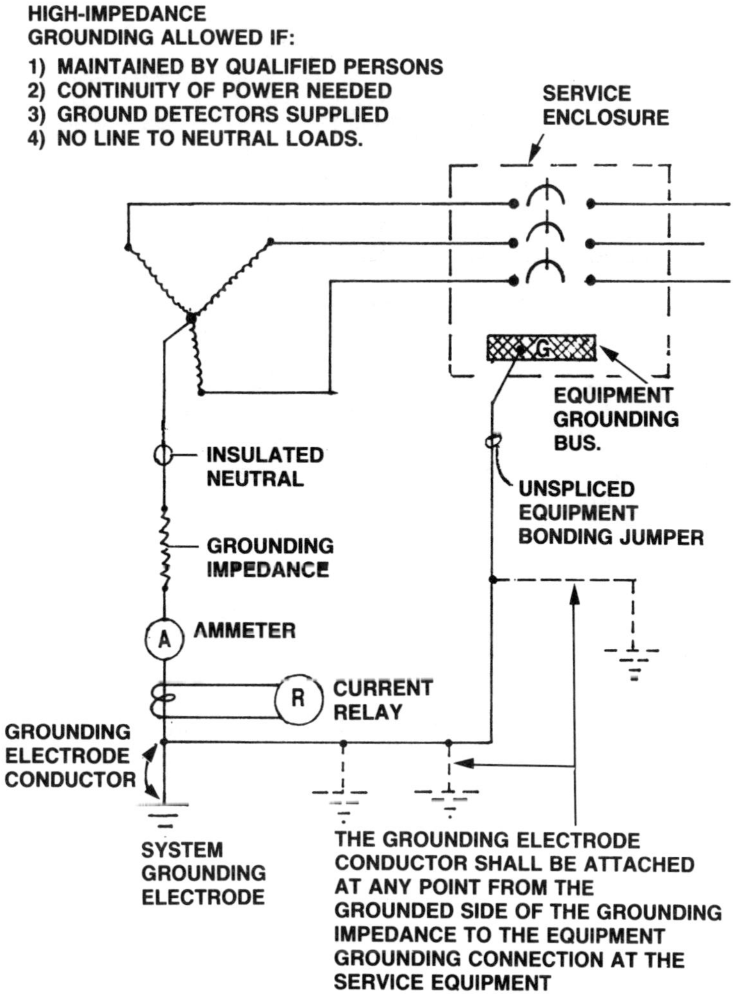

N.E.C. Sections 250-5 & 250-26

Grounding Of Separately Derived Systems

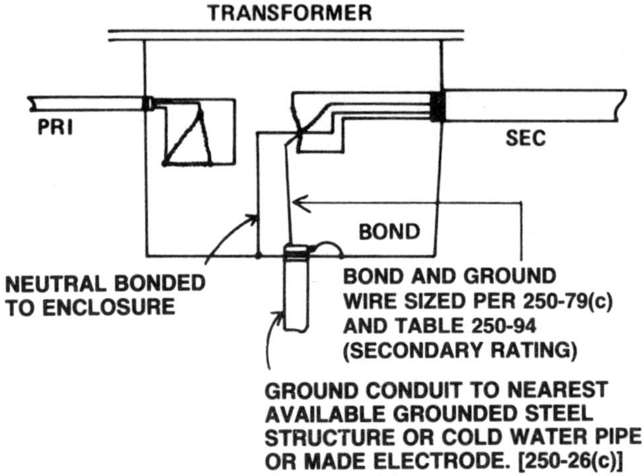

SYSTEMS HAVING A NEUTRAL ESTABLISHED SHALL BE GROUNDED AS REQUIRED FOR SERVICES.

N.E.C. Section 250-24 (a) X 2

Grounding At Several Buildings

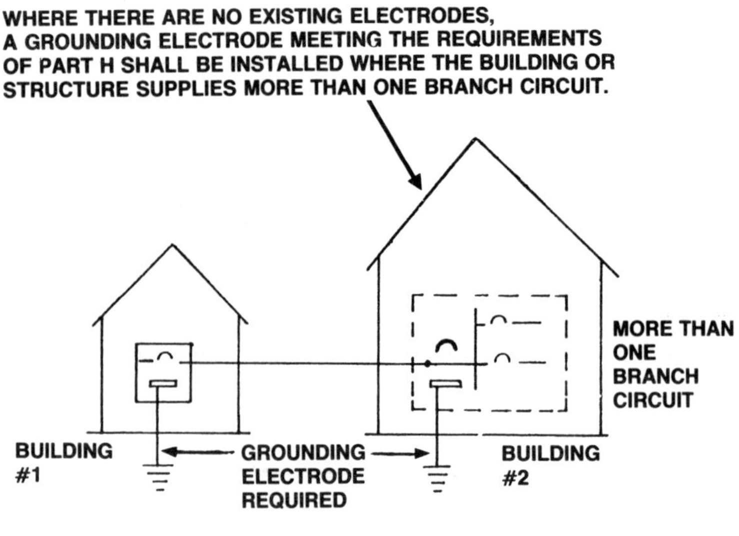

N.E.C. Section 250-26 (b), Exception #1

Grounding Of Enclosure

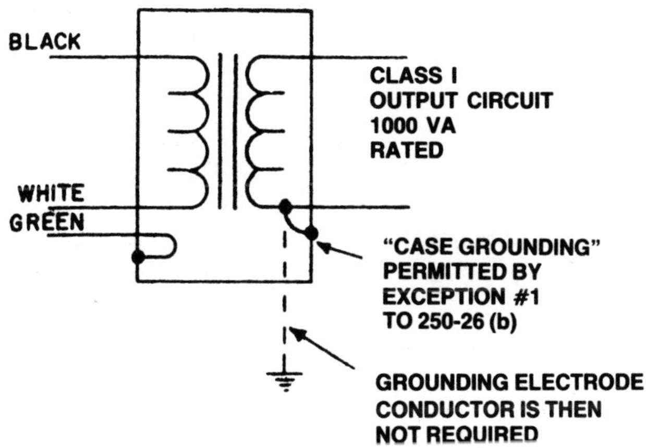

N.E.C. Section 250-42 (d)

Grounding Of Motors In Hazardous Areas

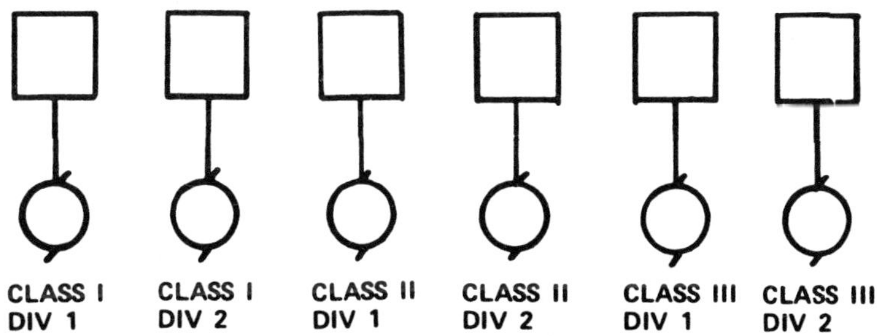

ALL EXPOSED METAL PARTS IN HAZARDOUS AREAS REQUIRE GROUNDING.

N.E.C. Section 250-45

Portable Electric Tools Require Grounding

PORTABLE ELECTRICAL TOOLS WHICH ARE HELD IN THE HAND WHILE BEING OPERATED ARE REQUIRED TO BE GROUNDED.

REGARDLESS OF VOLTAGE

EXCEPT LISTED TOOLS & APPLIANCES PROTECTED BY DOUBLE INSULATION.

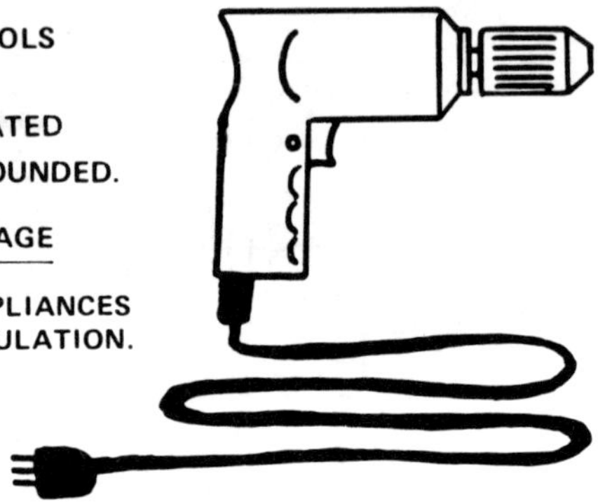

N.E.C. Section 250-54

Common Grounding Electrode

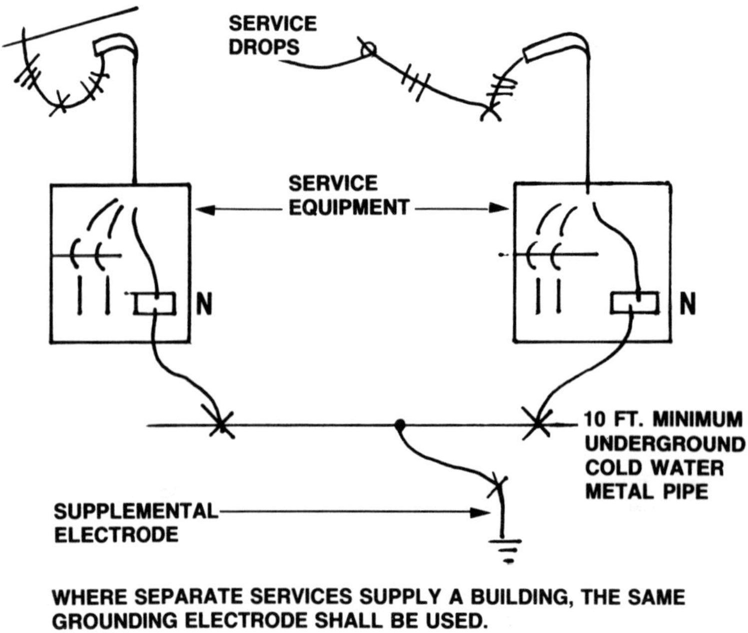

WHERE SEPARATE SERVICES SUPPLY A BUILDING, THE SAME GROUNDING ELECTRODE SHALL BE USED.

N.E.C. Section 250-59

Attachment Plugs

MOVABLE SELF RESTORING PRONG ATTACHMENT PLUG. LIMITED FOR USE ON HAND HELD EQUIPMENT ONLY.

FIXED PRONG ATTACHMENT PLUG

APPROVED

APPROVED

N.E.C. Section 250-59 (b)

Unapproved Methods Of Grounding Portable Equipment

NOT APPROVED, NO FIXED GROUND CONTACT.

NOT APPROVED, GROUND WIRE WRAPPED AROUND CORD. GROUND WIRE MUST BE A PART OF CORD ASSEMBLY.

N.E.C. Section 250-74

Approved Methods Of Grounding Receptacles To Boxes

WITH AN APPROVED JUMPER ATTACHED TO THE BOX WITH A #10 THREADED SCREW OR BOLT.

OR

WITH AN APPROVED JUMPER ATTACHED TO THE *BOX* WITH AN APPROVED CLIP. (NOT ATTACHED TO ANY BOX EXTENSION, PLASTER RING, ETC.)

NOTE:
APPROVED JUMPER IS AN INSULATED WIRE WITH A GREEN OR GREEN WITH YELLOW STRIPE ON THE INSULATION.

OR

THE RECEPTACLE ATTACHED TO THE BOX WITH APPROVED GROUNDING SCREWS.

OR

SELF-GROUNDING RECEPTACLES WHICH ARE ACCEPTABLE TO DEPARTMENT OF BUILDING AND SAFETY. (LABORATORY APPROVED).

N.E.C. Section 250-75

EFFECTIVE BONDING OF RACEWAY SYSTEM IS REQUIRED EVEN IF EQUIPMENT GROUNDING (BONDING) CONDUCTOR IS PRESENT.

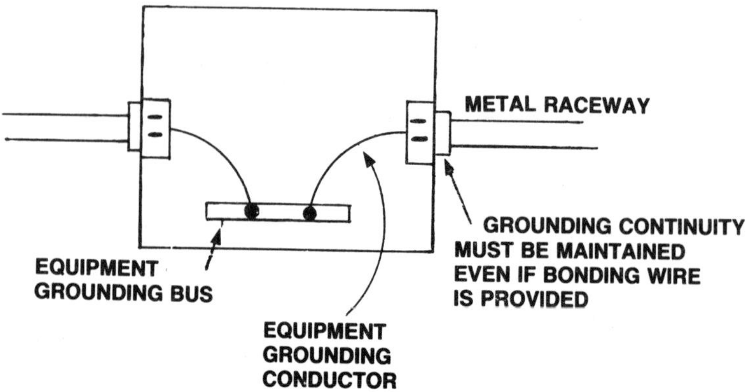

N.E.C. Section 250-79

Main & Equipment Bonding Jumpers

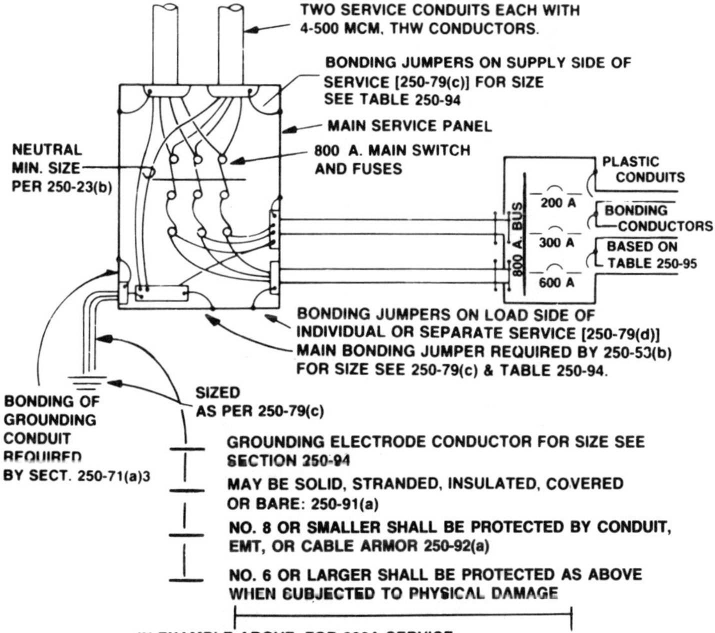

IN EXAMPLE ABOVE, FOR 800A SERVICE:

- BONDING JUMPER FOR GROUNDING CONDUIT: #2/0
- BONDING JUMPERS ON SUPPLY SIDE: #1/0
- BONDING JUMPERS ON LOAD SIDE: #1/0
- MAIN BONDING JUMPER: #2/0
- GROUNDING ELECTRODE CONDUCTOR: BASED ON 1000 MCM: #2/0
- EQUIPMENT BONDING CONDUCTORS FROM DISTRIBUTION BOARD:
  FOR 300A FEEDER: #6 FOR 300A FEEDER: #4 FOR 600A FEEDER: #1
- IF PARALLEL CONDUITS BETWEEN SERVICE EQUIPMENT AND DISTRIBUTION BOARD REQUIRE BONDING CONDUCTORS, EACH CONDUIT WILL REQUIRE A #1/0 CONDUCTOR SECT. 250-95

N.E.C. Section 250-79 (f)

Equipment Bonding Jumper

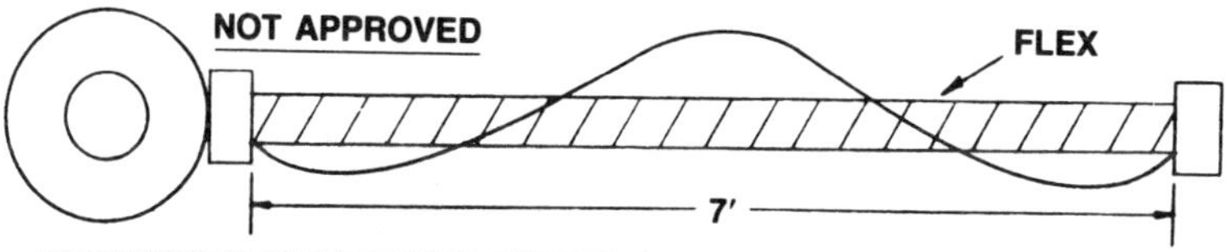

THE EQUIPMENT BONDING JUMPER OVER 6 FT. IN LENGTH SHALL BE INSTALLED INSIDE THE RACEWAY WITH THE CIRCUIT CONDUCTORS.

N.E.C. Section 250-80 (a) Ex.

Multi Occupancy Building

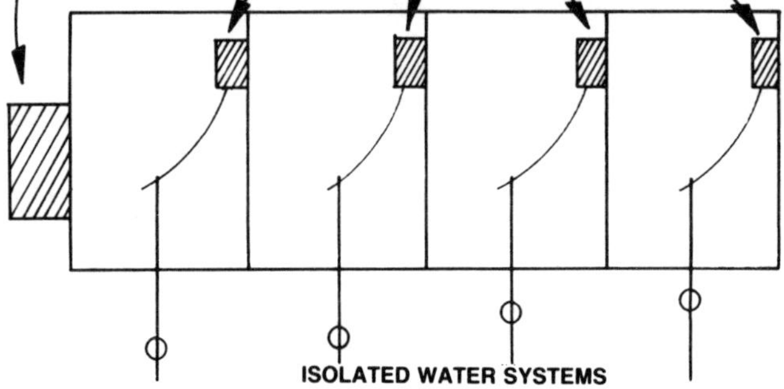

BONDING CONDUCTORS CONNECTED TO THE SUBPANEL AND SIZED AS PER TABLE 250-95

N.E.C. Section 250-80 (a) & (b)

Bonding Of Piping System

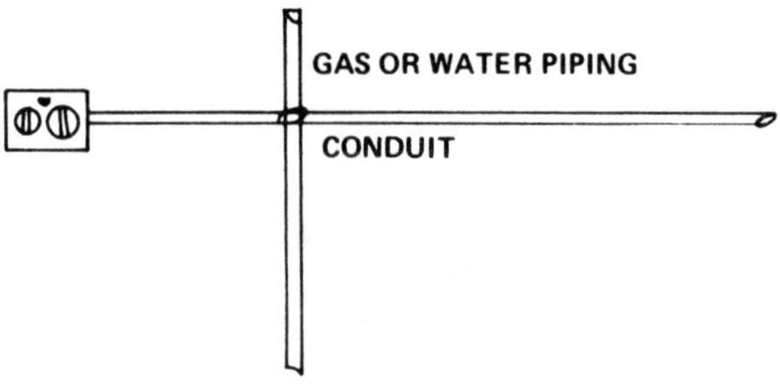

THE EQUIPMENT GROUNDING CONDUCTOR FOR THE CIRCUIT WHICH MAY ENERGIZE THE PIPING SHALL BE PERMITTED TO SERVE AS THE BONDING MEANS.

N.E.C. Section 250-91 (a), Ex. 2

Grounding Electrode Conductor

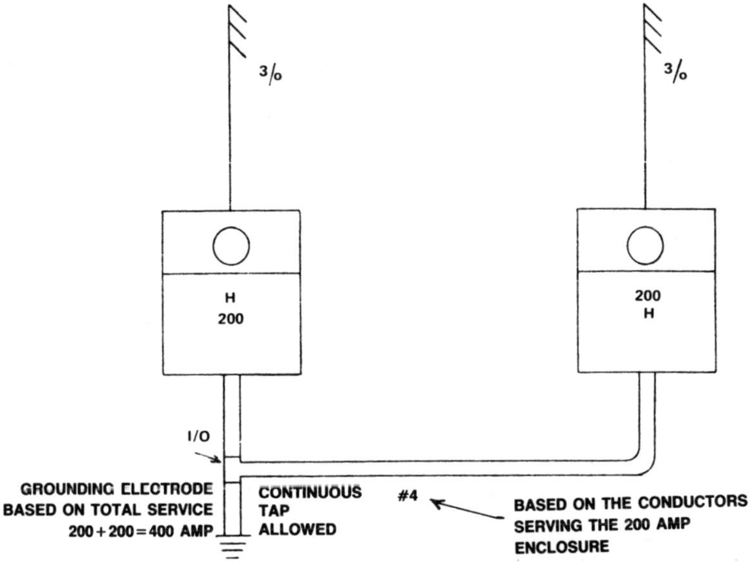

TAPS ARE PERMITTED TO THE GROUNDING ELECTRODE CONDUCTOR WHERE 2 TO 6 SERVICE DISCONNECTING MEANS ARE INSTALLED IN SEPARATE ENCLOSURES.

THE GROUNDING ELECTRODE CONDUCTOR SHALL BE SIZED IN ACCORDANCE WITH TABLE 250-94 BASED ON THE TOTAL SERVICE RATING. THE TAPS SHALL BE BASED ON SIZE OF THE SERVICE CONDUCTORS SERVING THE ENCLOSURE.

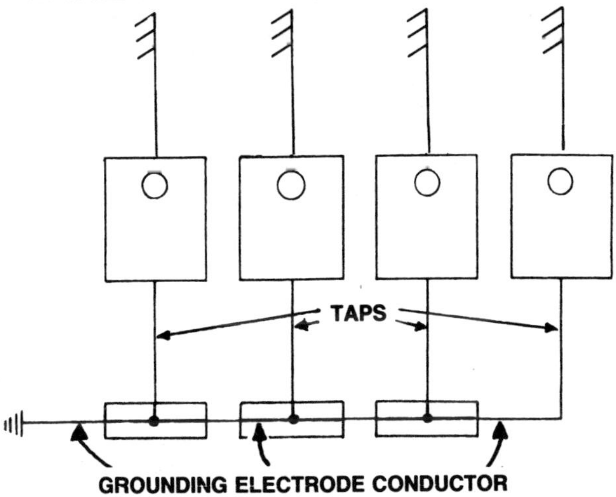

EXOTHERMIC WELDING PERMITTED IN INDUSTRIAL AND COMMERCIAL INSTALLATION.

N.E.C. Section 250-95

Equipment Grounding Conductors

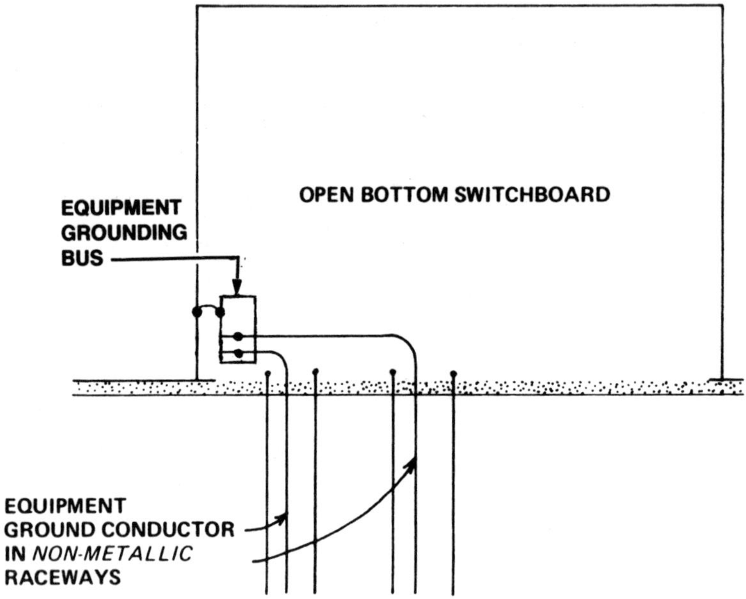

WHEN CONDUCTORS ARE RUN IN PARALLEL IN MULTIPLE RACEWAYS, A FULL SIZE EQUIPMENT GROUNDING CONDUCTOR SHALL ALSO BE RUN IN EACH RACEWAY. SEE SECTION 310-4.

N.E.C. Section 250-112

Ground Clamp Accessibility

EXCEPTION: A CONNECTION TO A DRIVEN, BURIED OR CONCRETE ENCASED ELECTRODE.

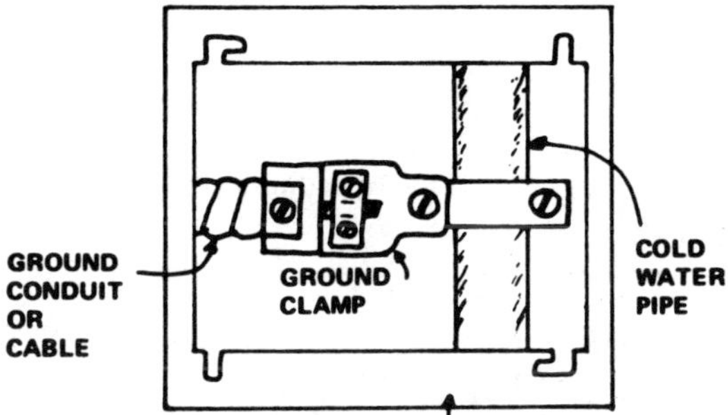

2 GANG SWITCH RING CENTERED SO CLAMP MAY BE REMOVED.

THE POINT OF CONNECTION OF THE GROUND CLAMP SHALL BE PERMANENTLY ACCESSIBLE FOR VISIBLE INSPECTION AND TEST.

NO SOLDER LUGS PERMITTED ON GROUND WIRES.

GROUND CLAMP SHALL NOT BE ATTACHED ON VALVES, FITTINGS, TEST NIPPLES, ETC.

N.E.C. Section 250-114 (a)

GROUNDING CONNECTIONS TO METAL BOXES MUST BE BY WAY OF U.L. LISTED GROUNDING DEVICES ON A DEDICATED GROUNDING SCREW.

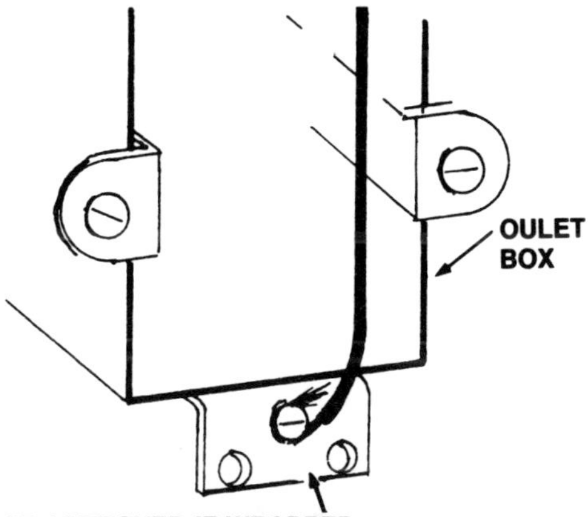

NOT APPROVED IF WRAPPED AROUND COVER MOUNTING SCREW

N.E.C. Section 300-3 (c)

Conductors Of Different Systems

(DEFINED 100-6 (h)

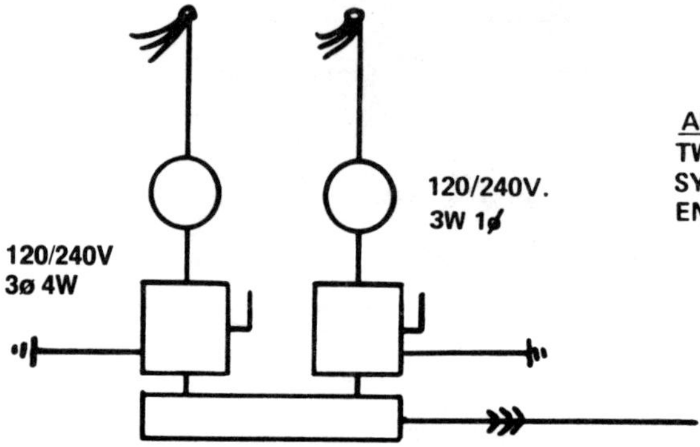

APPROVED
TWO DIFFERENT SYSTEMS IN SAME ENCLOSURE

N.E.C. Section 300-3 (c)

Conductors Of Different Systems In Same Enclosure

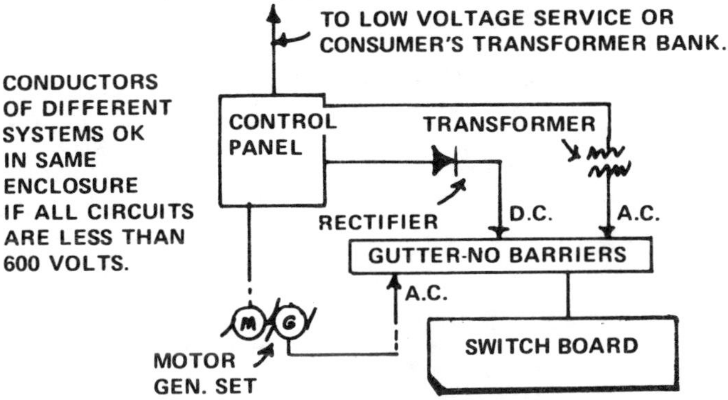

PROVIDED:
INSULATION OF *ALL* CONDUCTORS SHALL BE RATED FOR NOT LESS THAN THE HIGHEST VOLTAGE INVOLVED WITHIN THE RACEWAY.

N.E.C. Section 300-3 (c) 1

Conductors Of Different Systems

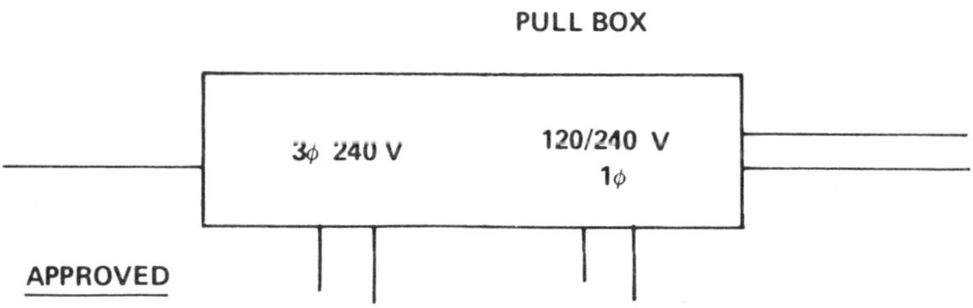

N.E.C. Section 300-3 (c) 2

Wires Of Different Systems Occupying Same Raceways, Boxes, Gutters, Cabinets, Etc.

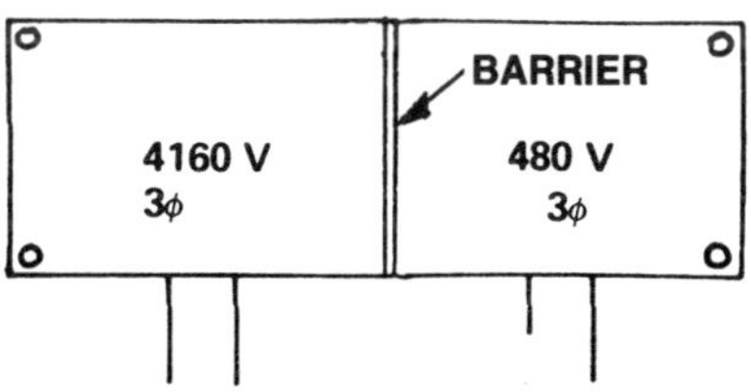

HIGH AND LOW VOLTAGE CONDUCTORS MUST BE SEPARATE.

N.E.C. Section 300-5 (c)

Direct Burial Wiring

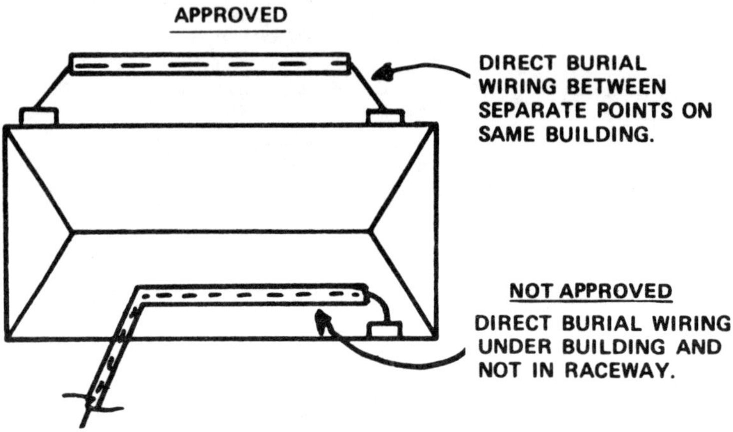

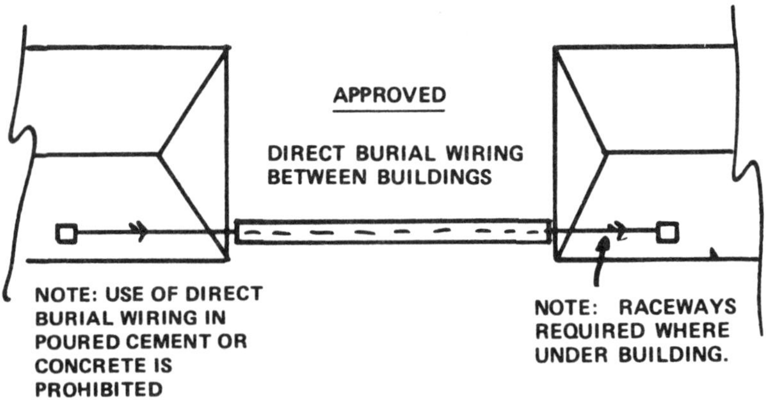

N.E.C. Section 300-5

Type U.S.E. Or U.F. Cable

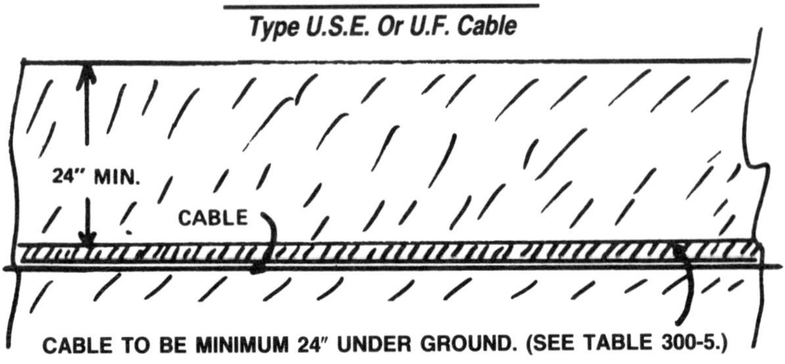

CABLE TO BE MINIMUM 24" UNDER GROUND. (SEE TABLE 300-5.)

N.E.C. Section 300-5 (a), Table 300-5

Class 2 Underground Wiring

DIRECT BURIAL INSTALLATION OF 30 VOLTS OR LESS.

CONDUCTORS AND CABLES SHALL BE BURIED AT LEAST 6 INCHES BELOW GROUND SURFACE.

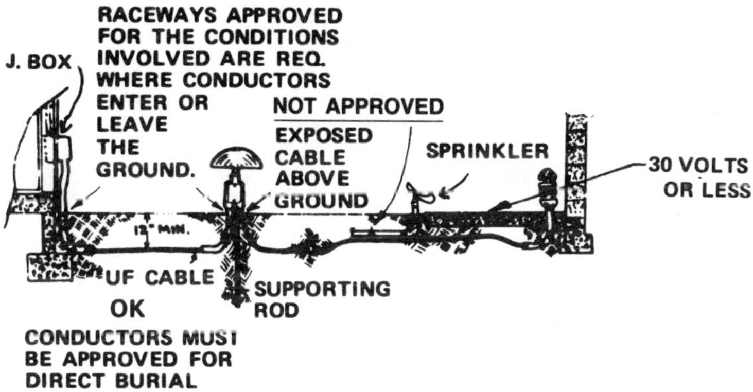

N.E.C. Sections 300-5 & 347-2 (g)

Rigid Non-metallic Conduit

P.V.C. RIGID NON METALLIC CONDUIT
FOR 600V AND LESS. SEE ARTICLE 710
FOR OVER 600 VOLTS.

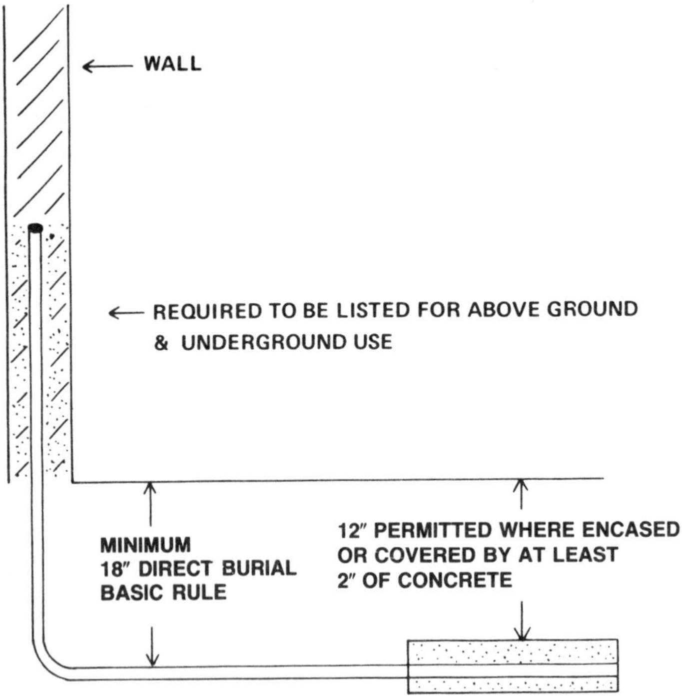

SEE TABLE 300-5

N.E.C. Section 300-5 (a)

Any Area Subject To Vehicular Traffic

SEE TABLE FOR OTHER APPLICATIONS

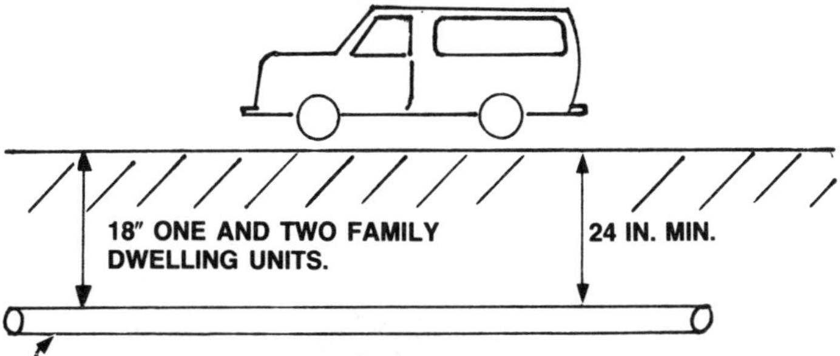

ANY WIRING METHOD, RIGID NON METALLIC, METAL CONDUIT, IMC, EVEN IF CONCRETE ENCASED.
EVEN ALLEYS, DRIVEWAYS, PARKING LOTS, ETC.

N.E.C. Section 300-6 (c)

Supporting Conduit In Indoor Wet Areas

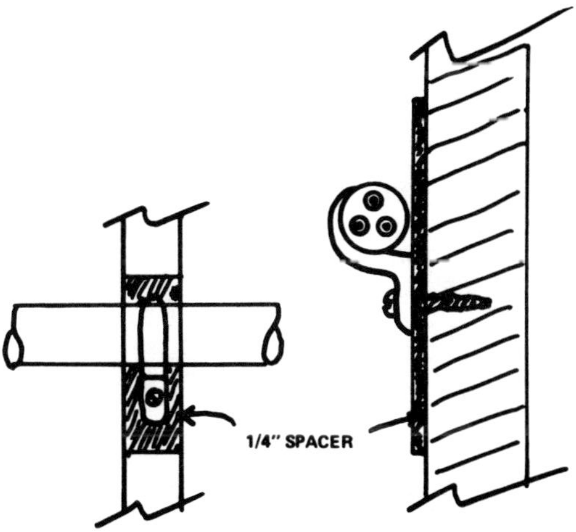

CONDUIT REQUIRED TO BE 1/4" FROM WALL IN WET LOCATIONS.

N.E.C. Sections 300-20 & 310-4

Induced Currents And Conductors In Parallel

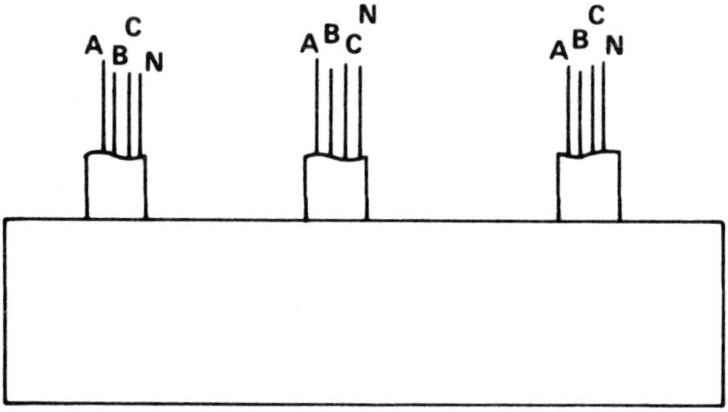

1. EACH CONDUIT, WHEN IN PARALLEL, SHALL CONTAIN EACH PHASE CONDUCTOR, THE NEUTRAL CONDUCTOR AND THE GROUND CONDUCTOR, IF USED. THE CONDUCTORS SHALL BE THE SAME SIZE, LENGTH. (EXCEPT THE GROUND).

2. THE CONDUCTORS SHALL BE AT LEAST 1/0 OR LARGER. 310-4 REFERS TO NOTE 8 OF TABLES 310-16 AND 310-18 REDUCTION FACTORS.

N.E.C. Section 300-21 (1)

Prevention Of Fire Spread

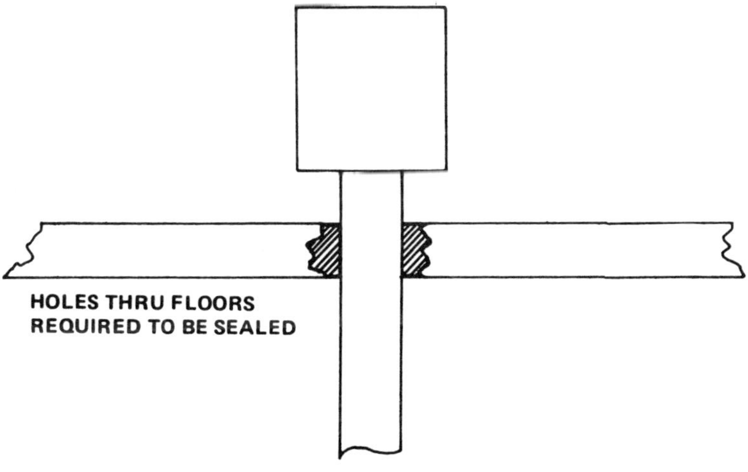

THE FIRE-PROTECTIVE RATING OF FIRE RATED WALLS, CEILINGS OR FLOORS WILL BE *SUBSTANTIALLY EQUIVALENT* TO ITS *ORIGINAL RATING.*

N.E.C. Section 300-22 (c)

Ceiling Heated Air Supply Plenum Chamber

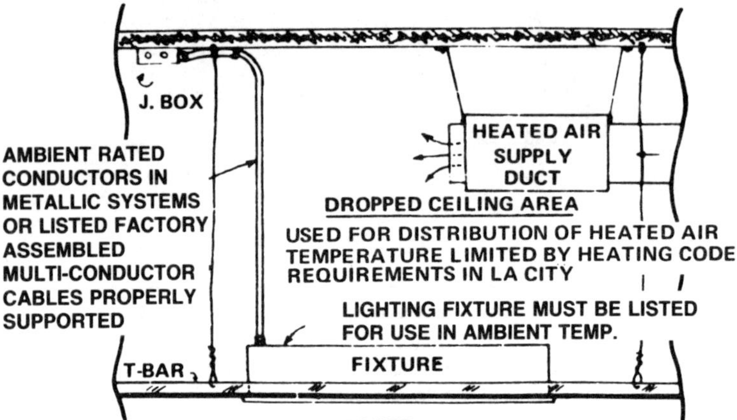

APPROVED

N.E.C. Section 300-22 (b)

Plenum Make-Up And Return Air Chamber

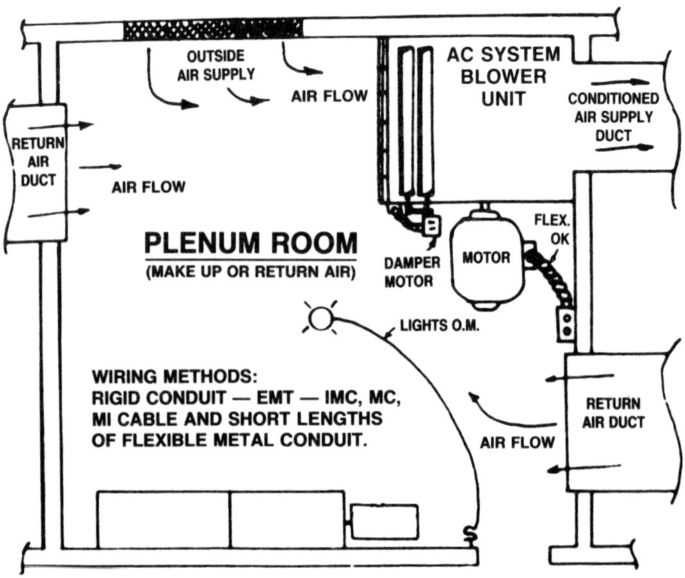

EQUIPMENT WITHIN PLENUM PERMITTED ONLY IF NECESSARY FOR DIRECT ACTION UPON THE CONTAINED AIR (MCCS OR CONTROL PANELS NOT PERMITTED).

N.E.C. Section 305-4

Temporary Wiring

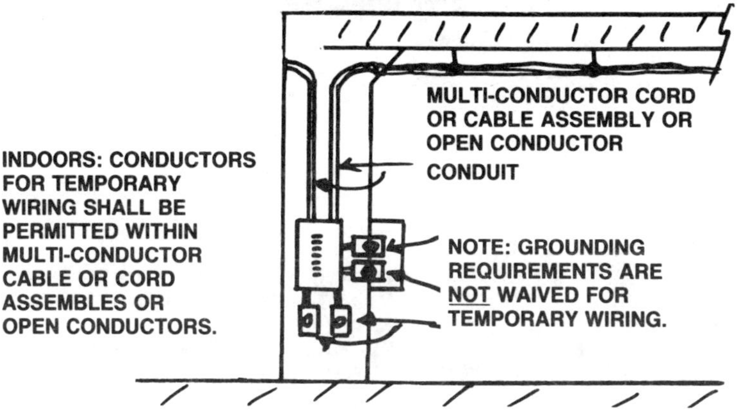

N.E.C. Section 305-6

Temporary Wiring

15 AND 20 AMPERE RECEPTACLE OUTLETS ON 1Ø CIRCUITS SHALL HAVE APPROVED GROUND FAULT CIRCUIT PROTECTION FOR PERSONNEL.

N.E.C. Section 310-4

Conductors In Parallel

1/0 ALUMINUM OR COPPER MINIMUM SIZE

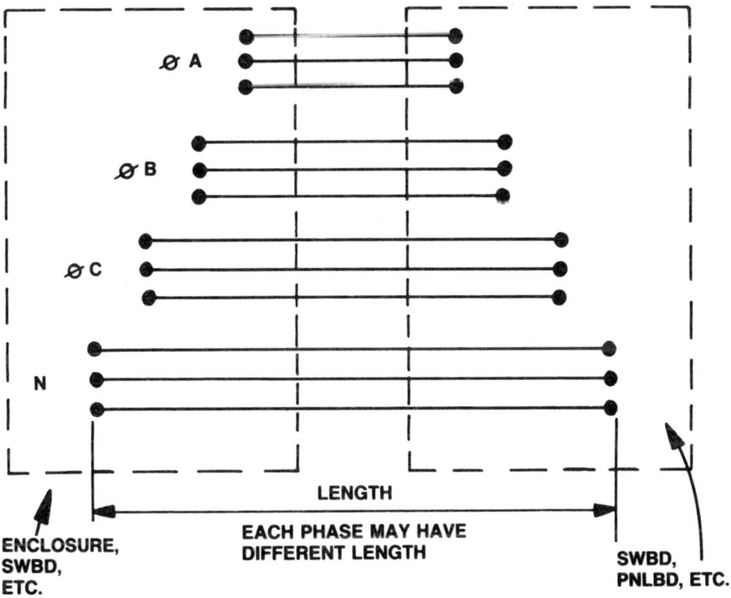

N.E.C. Section 310-15

Ampacity Reduction Factor

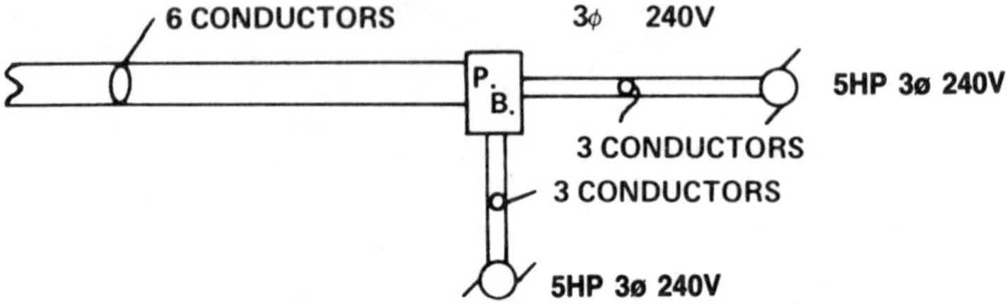

5HP = 15.2 AMPS X 125% = 19 AMPS (SEC. 430-22) #12 MIN. REQUIRED
6 CONDUCTORS IN RACEWAY AMPACITY LIMITED TO 80%
NOTE 8 TO TABLES 310-16 THROUGH 310-19
NO. 12 TW CONDUCTOR = 25 AMPERES X 80% AMPERES LOAD
THEREFORE NO. 12 CONDUCTORS ARE PERMITTED TO SUPPLY EACH MOTOR.

N.E.C. Section 310-15

Ampacity And Reduction Factors

(N.E.C. NOTE #10 OF NOTES TO N.E.C. TABLES 310-16 THROUGH 310-31)

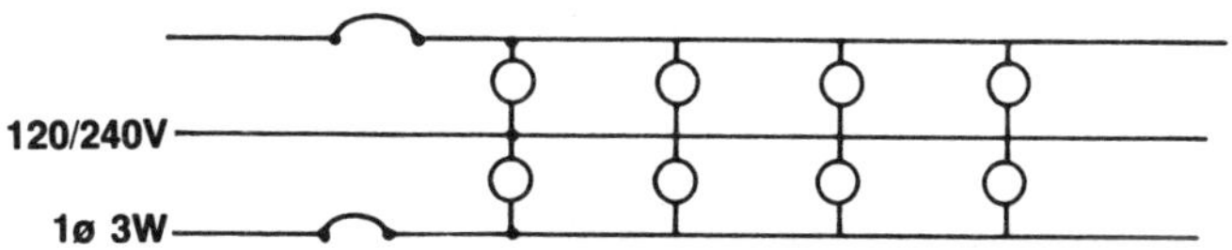

THE NEUTRAL CONDUCTOR WHICH CARRIES ONLY THE UNBALANCED CURRENTS FROM THE OTHER CONDUCTORS SHALL NOT BE COUNTED IN DETERMINING AMPACITIES AS PROVIDED IN NOTE 8.

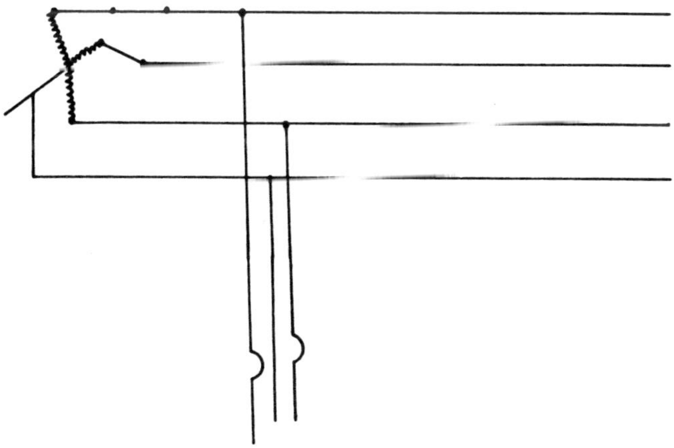

IN A 3 WIRE CIRCUIT CONSISTING OF TWO PHASE WIRES AND THE NEUTRAL OF A 4 WIRE, 3 PHASE WYE CONNECTED SYSTEM, A COMMON CONDUCTOR CARRIES THE SAME CURRENT AS THE OTHER CONDUCTORS AND SHALL BE COUNTED IN DETERMINING AMPACITIES AS PROVIDED IN NOTE 8, TABLES 310-16 THROUGH 310-31.

N.E.C. Section 310-15

Ampacity And Reduction Factors

(N.E.C. NOTE #8 OF NOTES TO N.E.C. TABLES 310-16 THROUGH 316-31)

AMPACITY REDUCTION FACTOR NOT REQUIRED
3Ø 4 WIRE

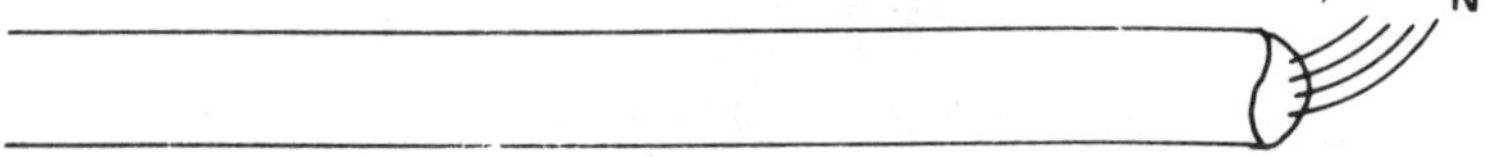

WHEN MORE THAN 3 CONDUCTORS ARE INSTALLED IN A RACEWAY, WHERE MAJOR PORTION IS NOT SUPPLYING ELECTRICAL DISCHARGE LIGHTING FIXTURES, THE AMPACITY FOR EACH CONDUCTOR SHALL BE IN ACCORDANCE WITH NOTE 8 TABLES 310-16 TO 310-18. THE NEUTRAL CONDUCTOR WHICH CARRIES THE UNBALANCED CURRENT FROM OTHER CONDUCTORS SHALL NOT BE COUNTED IN DETERMINING THE AMPACITY AS PROVIDED IN NOTES 8 & 10.

ACCORDANCE WITH NOTE 8 TABLES 310-16 TO 310-31. THE NEUTRAL CONDUCTOR WHICH CARRIES THE UNBALANCED CURRENT FROM OTHER CONDUCTORS SHALL NOT BE COUNTED IN AMPACITY AS PROVIDED IN NOTES 8 AND 10.

NOTE: CONDUCTORS MUST BE PROTECTED AT THE AMPACITY RATING AS SPECIFIED IN TABLES 310-16 THROUGH 310-31, AND AFTER DERATING ACCORDING TO NOTE #8 AND THE AMPACITY CORRECTION FACTORS WITHIN THE TABLES.

1990 NEC NOTE 8 TO TABLES 310–16 THROUGH 310–19 INCLUDES "ELECTRIC DISCHARGE LIGHTING, DATA PROCESSING, OR SIMILAR EQUIPMENT," WHICH CONTAIN HARMONIC CURRENTS AS LOADS WHICH REQUIRE THE NEUTRAL TO BE COUNTED AS A CURRENT CARRYING CONDUCTOR. THIS PAGE SHOULD BE MODIFIED TO REFLECT THIS NOTE.

N.E.C. Section 310-15 & 725-16 & 725-17

Ampacity And Reduction Factors

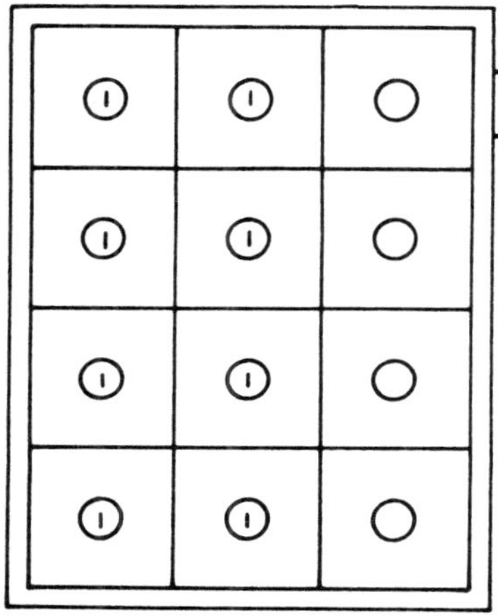

NUMBER OF CONTROL CONDUCTORS IN THE RACEWAY MAY BE DETERMINED BY TABLE 1 CHAPTER 9, CONDUCTOR AMPACITY NEED NOT BE DERATED IN ACCORDANCE WITH NOTE 8 TO TABLES 310-16 THROUGH 310-31.

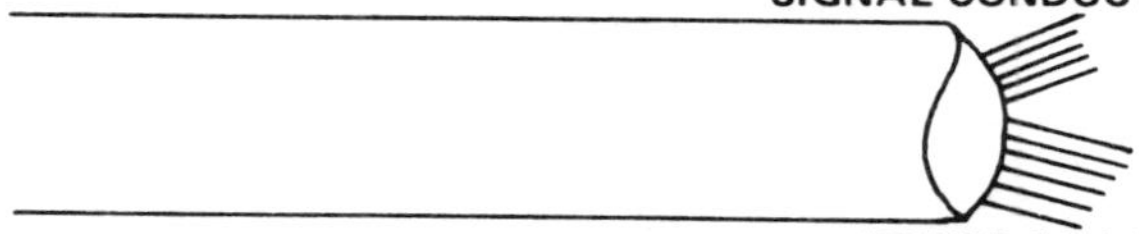

RACEWAY WITH REMOTE CONTROL CLASS ONE OR SIGNAL CONDUCTORS & POWER OR LIGHTING CONDUCTORS WHEN PERMITTED BY SEC. 725-15. THE AMPACITY OF CONDUCTORS AS REQUIRED BY NOTE 8 TO TABLES 310-16 THROUGH 310-31 SHALL BE DETERMINED BY THE NUMBER OF POWER OR LIGHTING CONDUCTORS IN THE RACEWAY.

POWER SUPPLY AND CLASS I CONDUCTORS SHALL BE PERMITTED ONLY WHERE THE EQUIPMENT POWERED IS FUNCTIONALLY ASSOCIATED.

N.E.C. Section 310-15

Ampacity And Reduction Factors

(N.E.C. NOTE #10 OF NOTES TO N.E.C. TABLES 310-16 THROUGH 310-31)

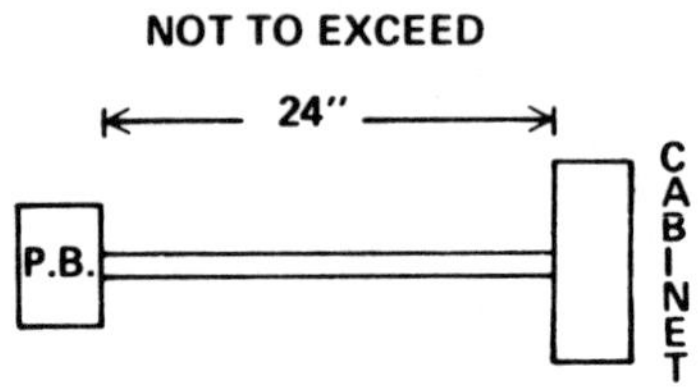

CONDUIT NIPPLES NOT OVER 24" IN LENGTH MAY BE FILLED TO 60% OF ITS TOTAL CROSS-SECTIONAL AREA. REDUCED AMPACITY FACTOR IN ACCORDANCE WITH NOTE 8 OF TABLE 310-16 THROUGH 310-31 DOES NOT APPLY TO THIS CONDITION.

N.E.C. Section 333-5

A/C CABLE INSTALLED IN INSULATED WALLS, CEILINGS AND ATTICS SHALL HAVE AN INSULATION RATING OF 90°C WITH AN AMPACITY OF 60°C.

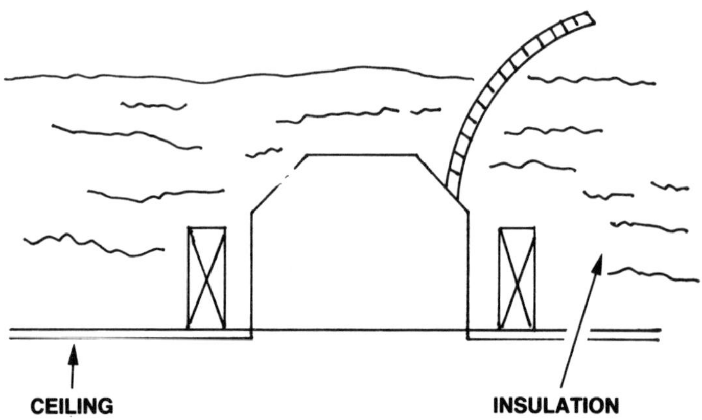

N.E.C. Section 333-7

Unsupported Lengths Of Type AC Cable

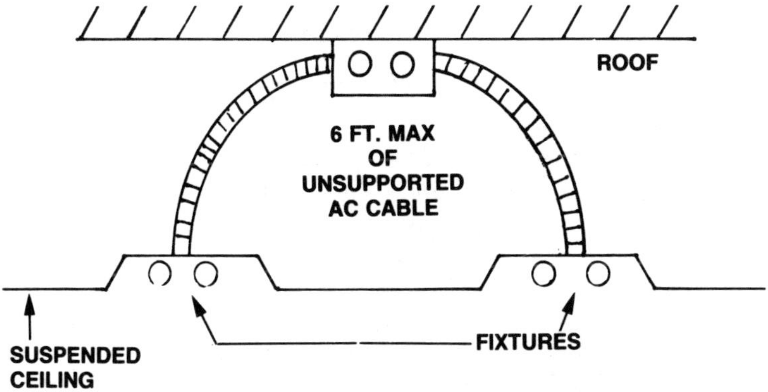

N.E.C. Section 346-9 (b)

Running Threads

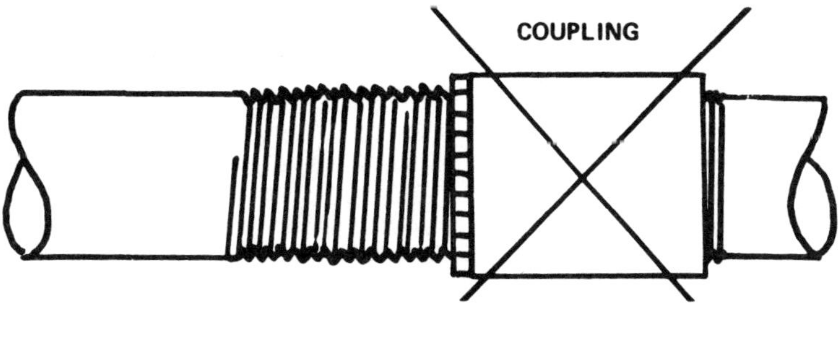

RUNNING THREADS SHALL NOT BE USED ON CONDUIT FOR CONNECTIONS AT COUPLINGS.

N.E.C. Section 347-2 & 3

Installation Of Rigid Non-metallic Conduit

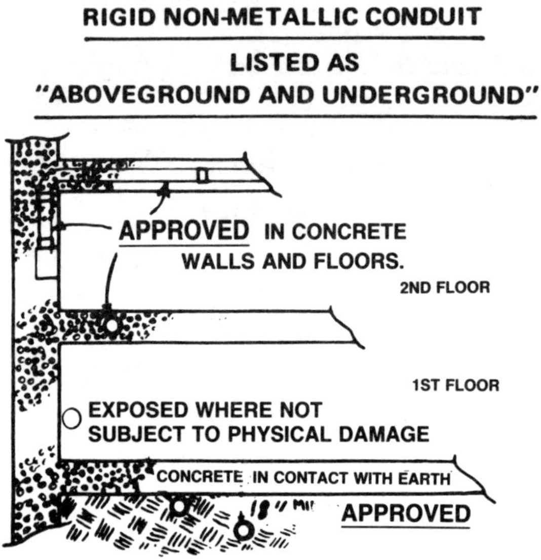

N.E.C. Section 347-2 (f)

Installation Of Rigid Non-metallic Conduit

RIGID NON-METALLIC CONDUIT

LISTED AS

"ABOVEGROUND AND UNDERGROUND"

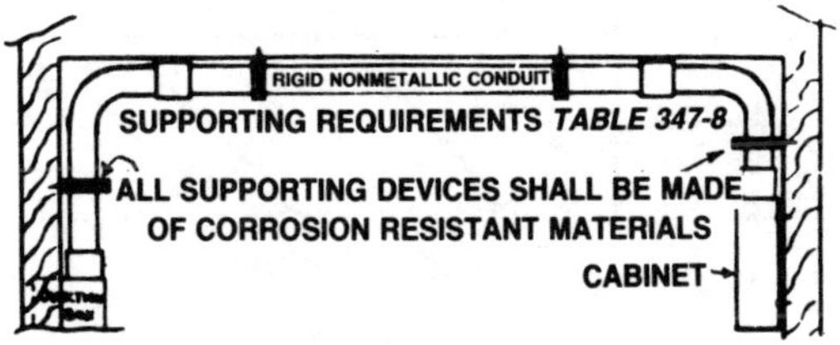

CAN BE RUN ON THE SURFACE IN *INDOOR LOCATIONS* EXPOSED TO SEVERE CORROSIVE CONDITIONS OR AREAS FREQUENTLY WASHED OR CONTAINING EXCESSIVE MOISTURE OR WHERE NOT SUBJECT TO PHYSICAL DAMAGE.

N.E.C. Section 348-12

Supporting EMT

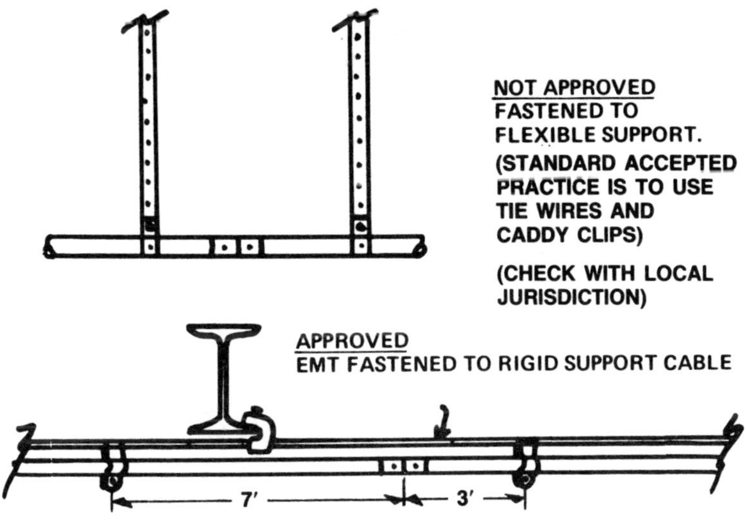

EMT SHALL BE SECURELY FASTENED IN PLACE WITHIN 3′ OF EACH CONNECTOR AND WITHIN 3′ OF EACH FITTING.

N.E.C. Section 350-2 — 225-10

Flexible Conduit On The Exterior Of Buildings

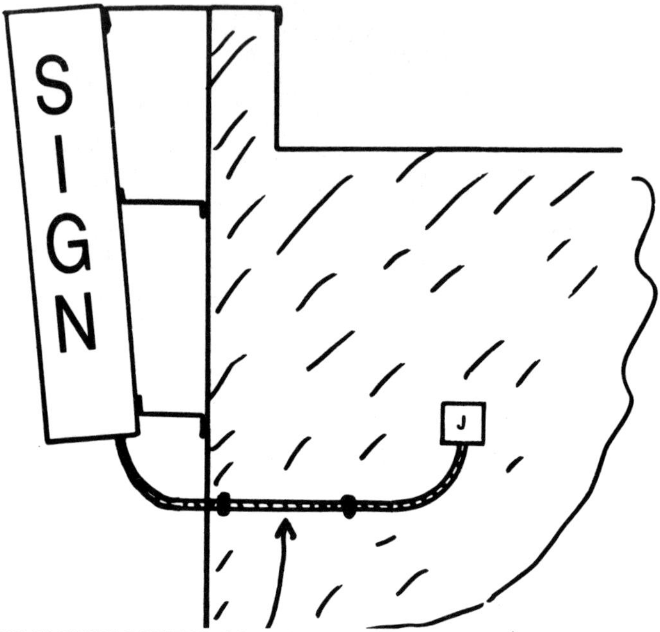

WHERE FLEXIBLE METAL CONDUIT IS INSTALLED ON THE EXTERIOR OF BUILDINGS IT SHALL:

1. HAVE CONDUCTORS SUITABLE FOR WET LOCATIONS AND
2. BE INSTALLED TO PRECLUDE THE ENTRANCE OF WATER IN ANY ENCLOSURE.

NOTE: THIS IS NOT INTENDED TO APPLY TO LIQUID -TIGHT FLEXIBLE METAL CONDUIT USED IN ACCORDANCE WITH THE PROVISIONS OF ART. 351.

N.E.C. Section 350-4

Fastening Flexible Conduit

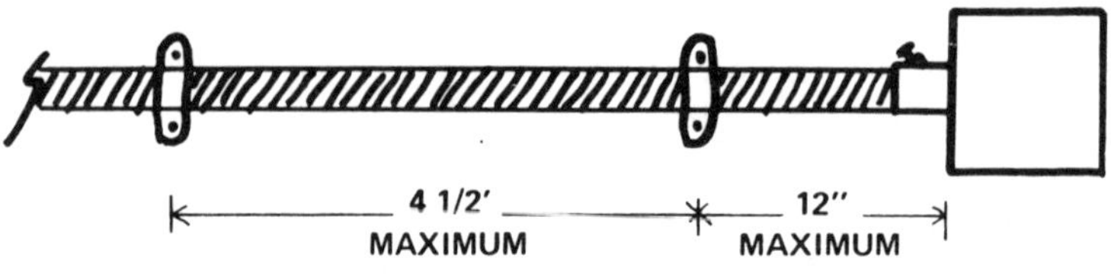

N.E.C. Section 350-5, Ex. #1

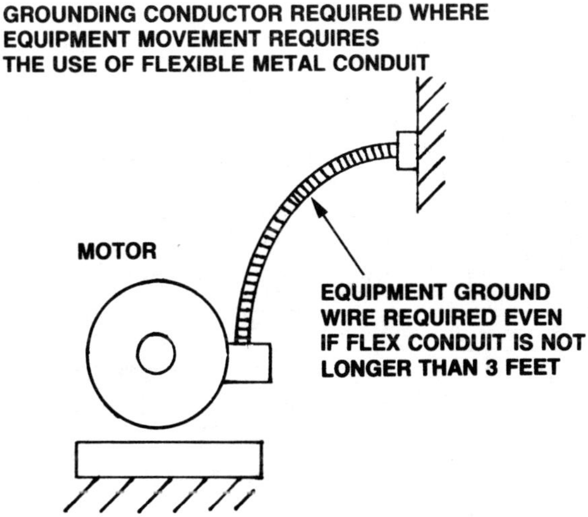

N.E.C. Sections 354-6, 356-6 & 358-12

Splices In Underfloor And Cellular Raceways

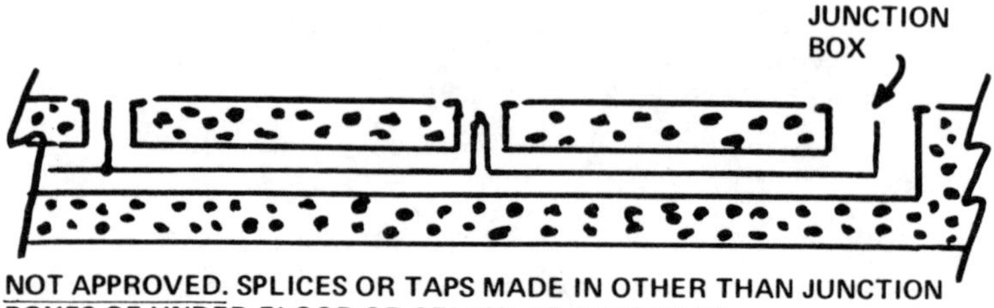

NOT APPROVED. SPLICES OR TAPS MADE IN OTHER THAN JUNCTION BOXES OF UNDER FLOOR OR CELLULAR RACEWAYS.

NOTE: LOOPING OF CONDUCTORS IS NOT CONSIDERED AS A SPLICE

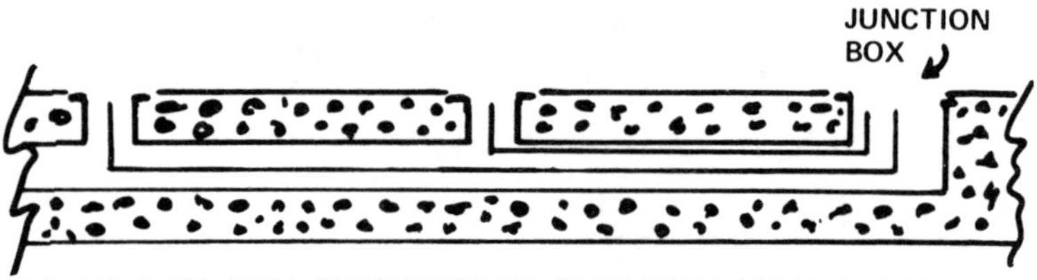

O.K. SEPARATE WIRE EXTENDING FROM JUNCTION BOX TO EACH RECEPTACLE.

N.E.C. Sections 354, 356 & 358

Underfloor Raceways

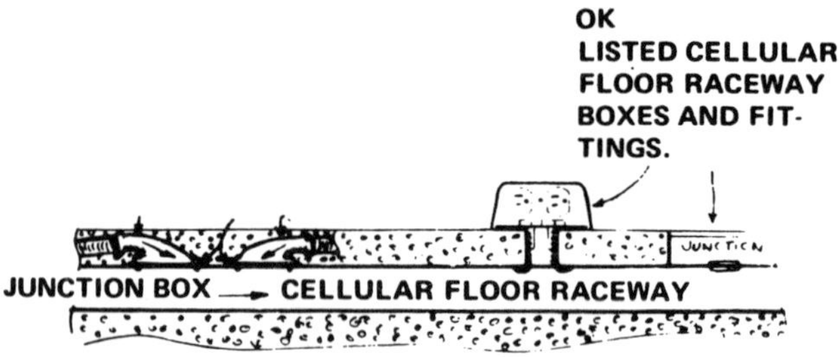

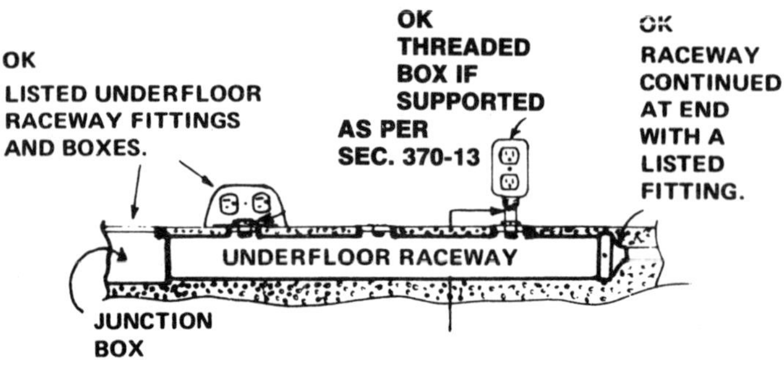

N.E.C. Section 362-5

Rating Of Conductors In Auxiliary Gutter

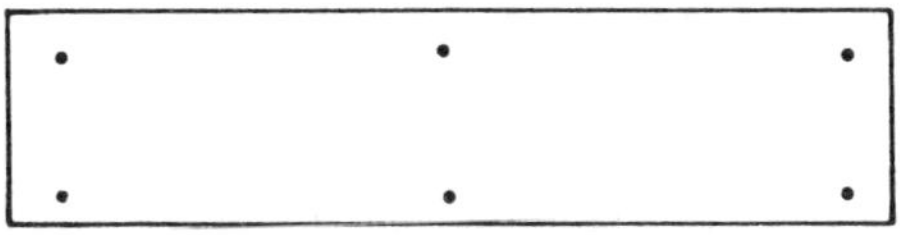

AUXILIARY GUTTER WITH NOT OVER 30 BRANCH CIRCUIT CONDUCTORS AND NOT OVER 20% FILL NEED NOT BE DERATED.

N.E.C. Section 362-5, Exception 1

AUXILIARY GUTTER WITH MORE THAN 30 CONDUCTORS AND NOT OVER 20% FILL SHALL BE DERATED ACCORDING TO NOTES TO TABLES 310-16 THROUGH 310-19 NOTE 8.

N.E.C. Section 364-11

Busway Reduced In Size

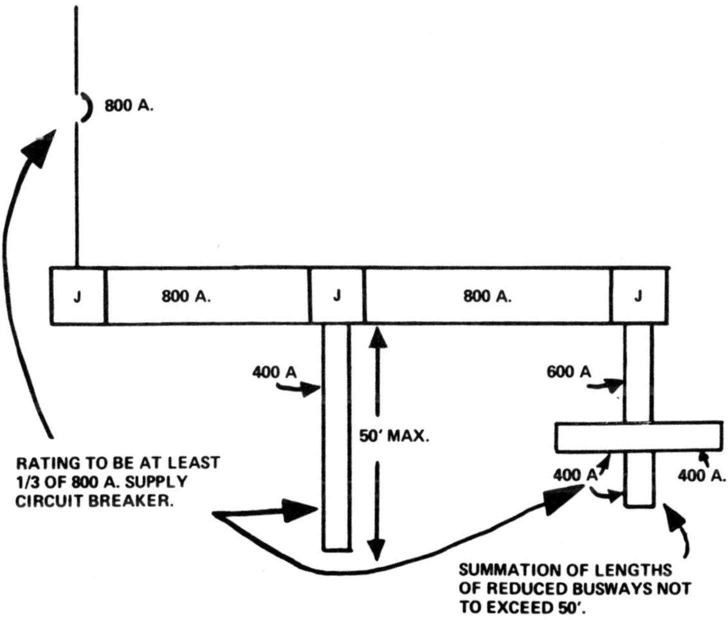

N.E.C. Section 364-11

Busway Reduced In Size

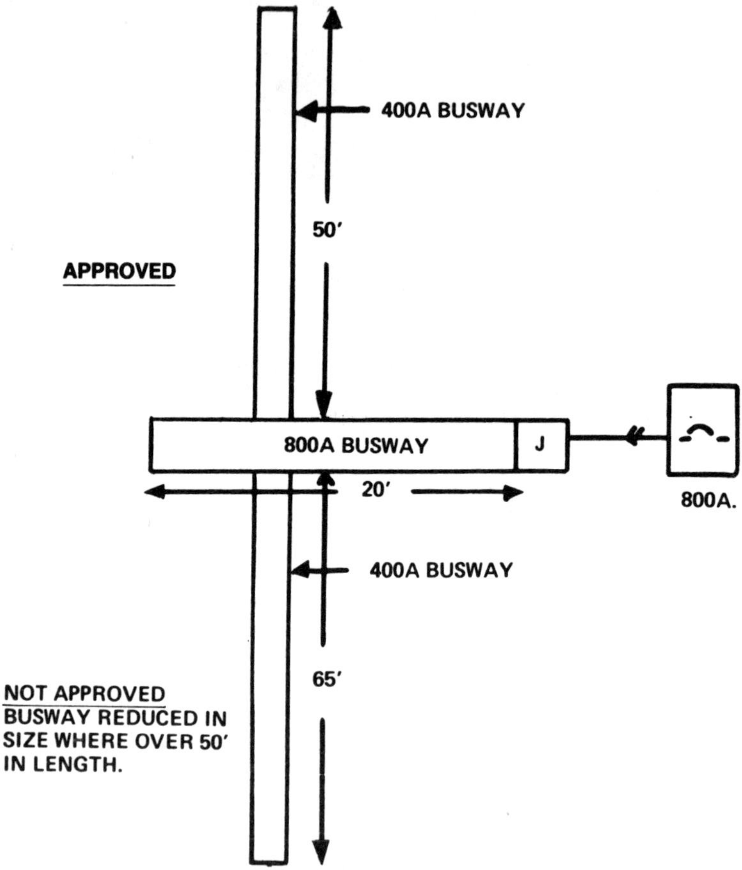

N.E.C. Section 364-12

Busway Branch Circuits

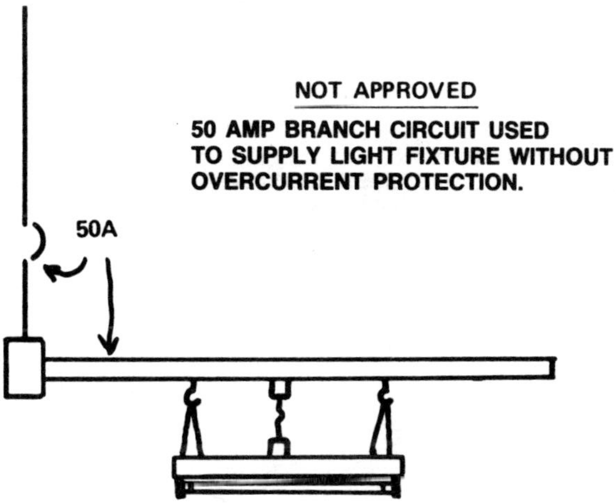

N.E.C. Section 364-13

Busway 20 Ampere Branch Circuit Conforming To The Provisions Of Articles 210 & 410

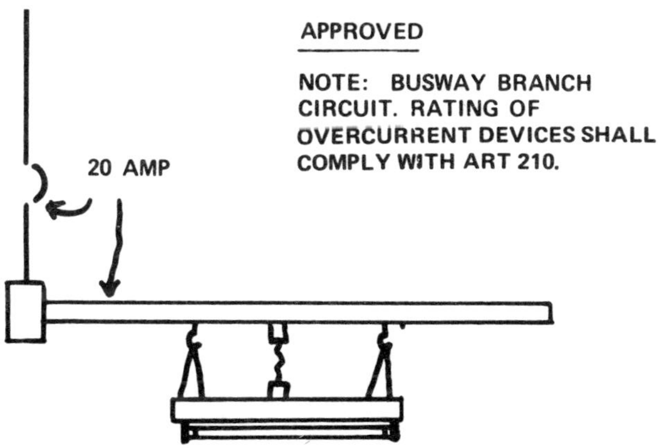

N.E.C. Section 364-25

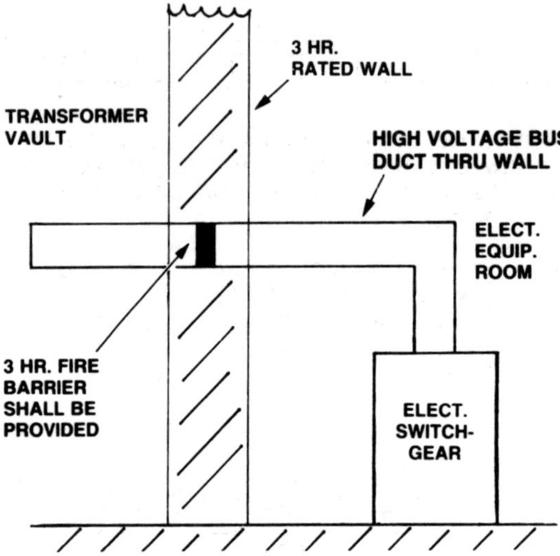

N.E.C. Section 370-3

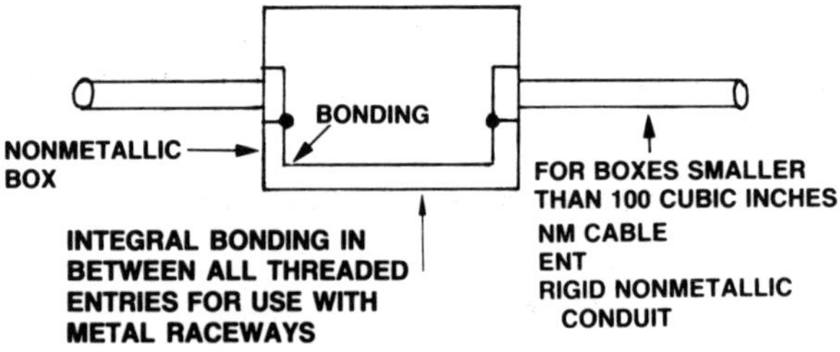

N.E.C. Section 370-6 (b) & (c)

Conduit Bodies

WHERE THE CONDUCTORS ARE SPLICED THE MAXIMUM NUMBER OF CONDUCTORS PERMITTED ARE COMPUTED USING THE VOLUME LISTED IN TABLE 370-6(a).

WHERE THE CONDUCTORS ARE NOT SPLICED THE MAXIMUM NUMBER OF CONDUCTORS PERMITTED SHALL BE BASED IN ACCORDANCE WITH THE NUMBER OF CONDUCTORS PERMITTED IN THE CONDUIT IT IS ATTACHED.

N.E.C. Section 370-16

Combination of Conductors In Box

VOLUME REQUIRED PER CONDUCTOR	BOX ALLOWABLE CAPACITY
No. 14 = 2 Cu. In. Ea.	4/0 X 1 1/2 Deep = 15.5 Cu.In.
No. 12 = 2.25 Cu. In. Ea.	4/S X 1 1/2 D = 21.0 Cu.In.
No. 10 = 2.5 Cu. In. Ea.	4/S X 2 1/8 D = 30.3 Cu.In.
No. 8 = 3 Cu. In. Ea.	5/S X 1 1/2 D = 29.5 Cu.In.
No. 6 = 5 Cu. In. Ea.	5/S X 2 1/8 D = 42.0 Cu.In.

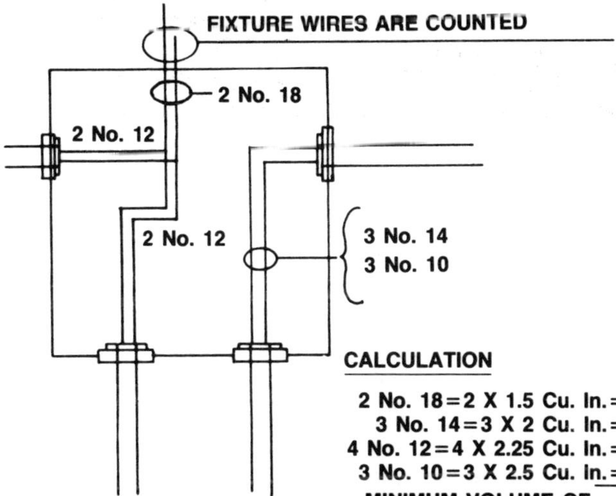

CALCULATION

2 No. 18 = 2 X 1.5 Cu. In. = 3
3 No. 14 = 3 X 2 Cu. In. = 6
4 No. 12 = 4 X 2.25 Cu. In. = 9
3 No. 10 = 3 X 2.5 Cu. In. = 7.5

MINIMUM VOLUME OF BOX WITH RING 25.5

N.E.C. Section 370-17

Non-metallic Boxes

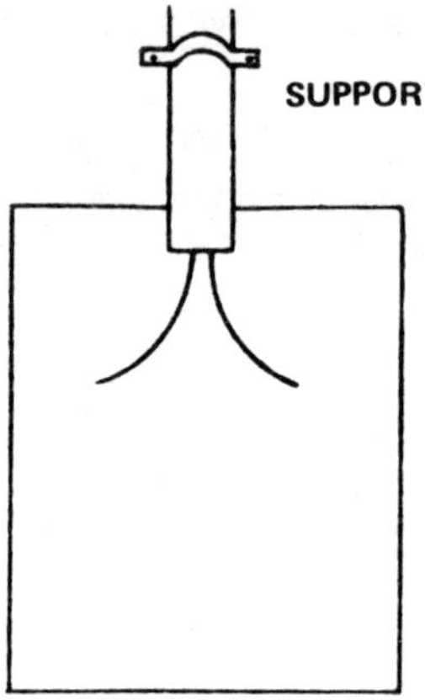

SUPPORT WITH IN 8" OF BOX.

APPROVED

NONMETALLIC SHEATHED CABLE PERMITTED TO ENTER SINGLE GANG BOX NOT LESS THAN ¼" WITHOUT CLAMP.

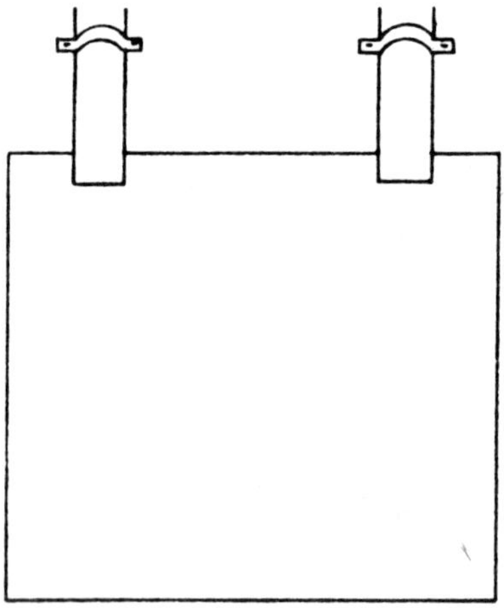

NOT APPROVED

NONMETALLIC SHEATHED CABLE ENTERING OTHER THAN SINGLE GANG BOXES WITHOUT BEING SECURED TO BOX.

EXCEPTION: WHERE NONMETALLIC-SHEATHED CABLE IS USED WITH BOXES NO LARGER THAN A NOMINAL SIZE 2¼ INCH BY 4 INCH MOUNTED IN WALLS AND WHERE THE CABLE IS FASTENED WITHIN 8 INCHES OF THE BOX MEASURED ALONG THE SHEATH AND WHERE THE SHEATH EXTENDS INTO THE BOX NO LESS THAN ¼ INCH (6.35MM), SECURING THE CABLE TO THE BOX SHALL NOT BE REQUIRED.

N.E.C. Section 370-20

Boxes In Wall Or Ceiling

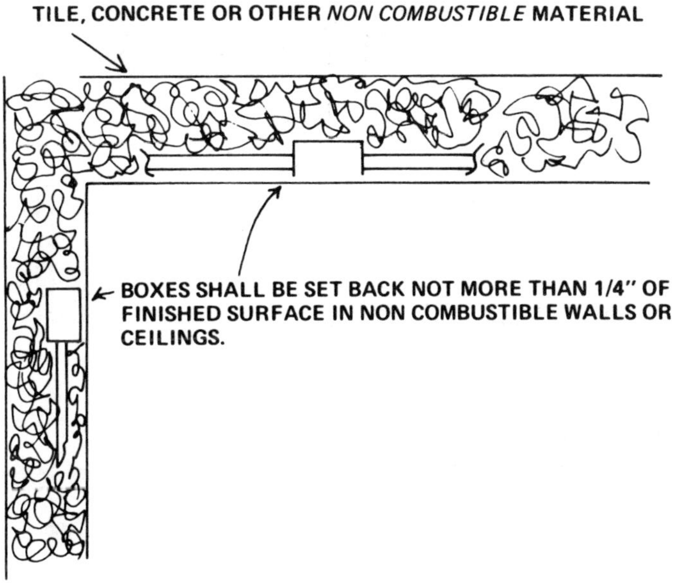

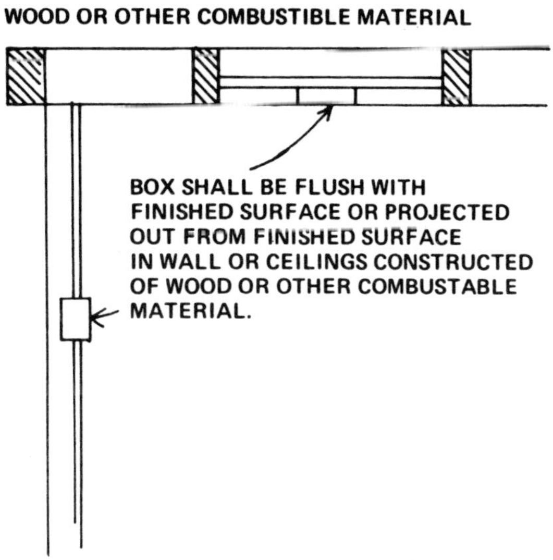

N.E.C. Section 370-21

Repair Of Plaster And Drywall Or Plasterboard

BROKEN PLASTER AT EDGE OF BOX. HOLE REQUIRED TO BE PATCHED. FIRE COULD ENTER WALL SPACE

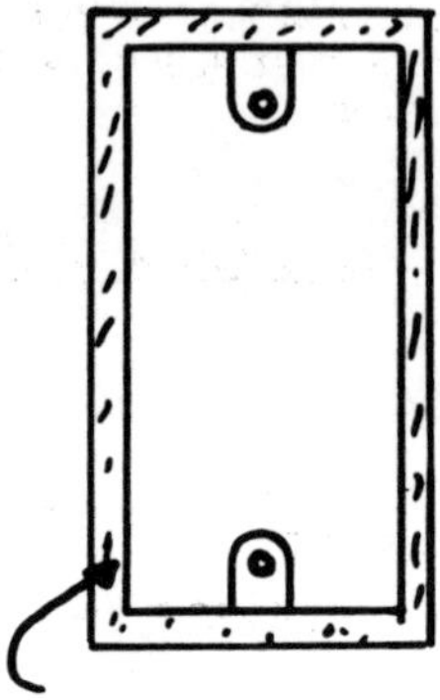

THERE SHALL BE NO GAP BETWEEN BOX AND DRYWALL GREATER THAN 1/8 INCH.
SPACE IS REQUIRED TO BE FILLED, ELSE FIRE COULD ENTER WALL SPACE.

N.E.C. Section 370-15 (b)

Covering Combustible Ceilings At Canopies

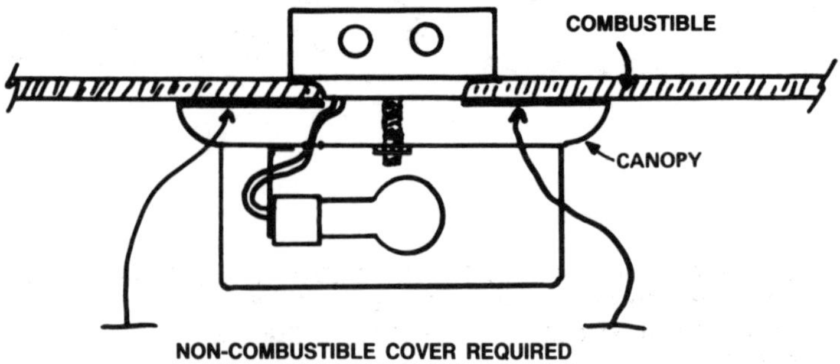

NON-COMBUSTIBLE COVER REQUIRED BETWEEN CANOPY AND COMBUSTIBLE MATERIALS.

N.E.C. Section 370-17 (c)

Boxes At Fan Outlets

A CEILING PADDLE FAN MUST NOT BE SUPPORTED FROMA CEILIN OUTLET BOX – UNLESS THE BOX IS LISTED FOR THE PURPOSE SEE 422–18

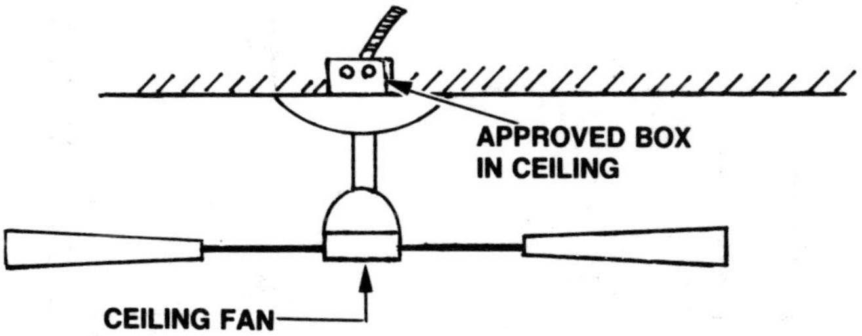

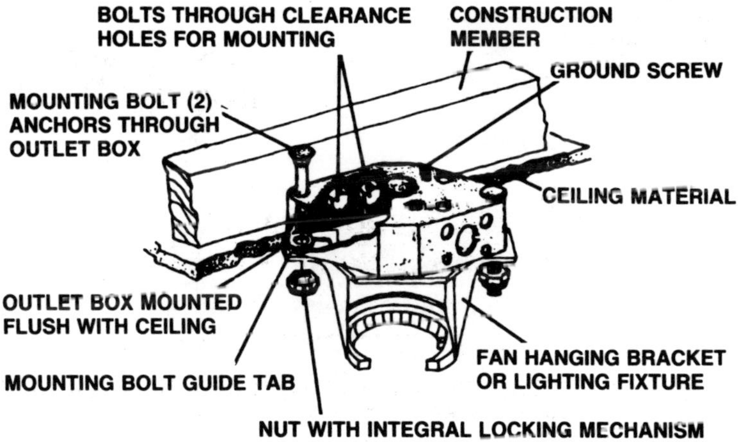

- REINFORCED OUTLET BOX BOLTS SOLIDLY TO JOISTS — MOUNTS EITHER THROUGH THE SIDE OR THE TOP OF BOX.
- SUITABLE FOR ROMEX CABLE, FLEXIBLE CONDUIT, EMT, OR RIGID.
- SUITABLE FOR WIRE-THROUGH APPLICATIONS.

N.E.C. Section 370-18

Conduit Bodies

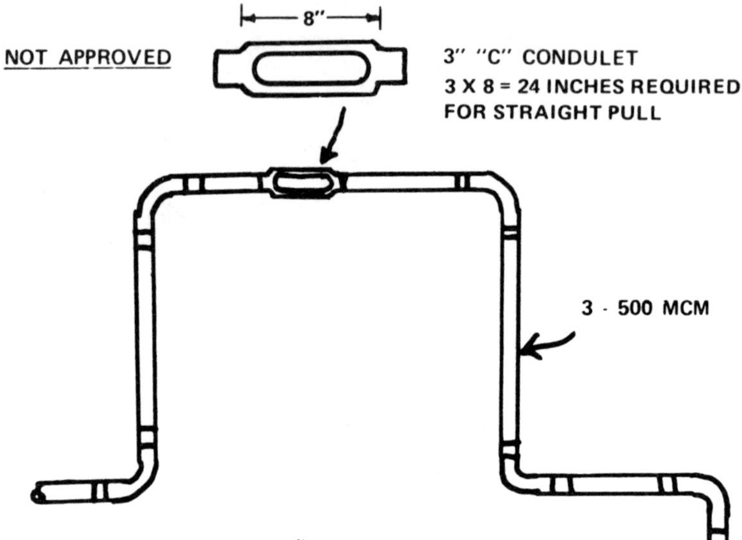

CONDUIT BODIES FOR #4 OR LARGER WIRES SHALL BE SIZED THE SAME AS BOXES CONTAINING #4 OR LARGER WIRES.
EXCEPT WHERE MARKED PERMANENTLY FOR SIZE AND NUMBER OF CONDUCTORS.

N.E.C. Section 370-28

Wireway Or Auxiliary Gutter Used As Junction Box

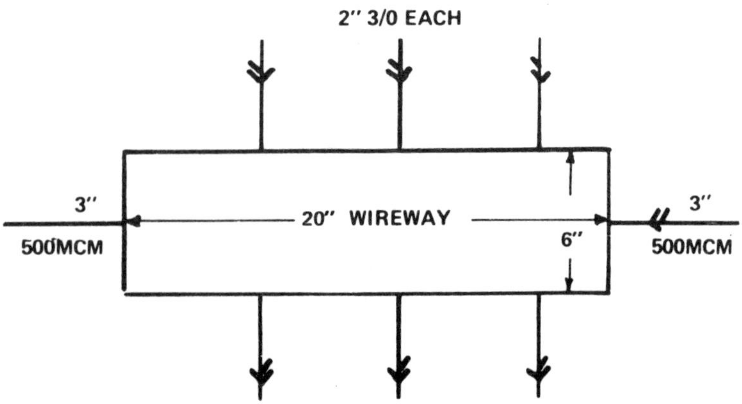

NOT APPROVED

WIREWAY OR AUXILIARY GUTTER USED IN LIEU OF A JUNCTION BOX NOT HAVING REQUIRED DIMENSIONS.

N.E.C. Section 373-2

Panelboard Spacing In Wet Locations

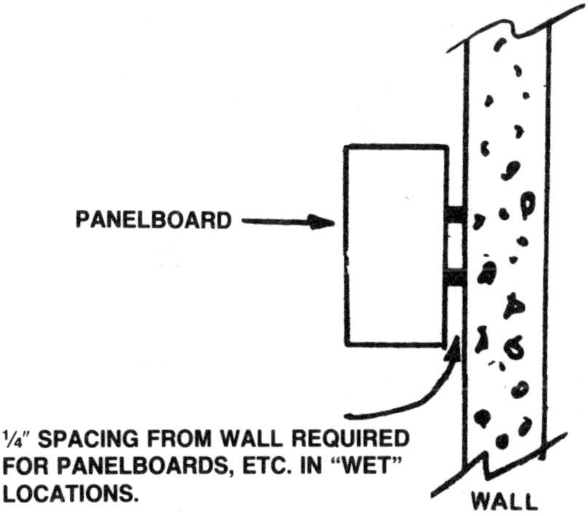

N.E.C. Sections 373-6 & 373-7

Deflecting Conductors In Panel Boards

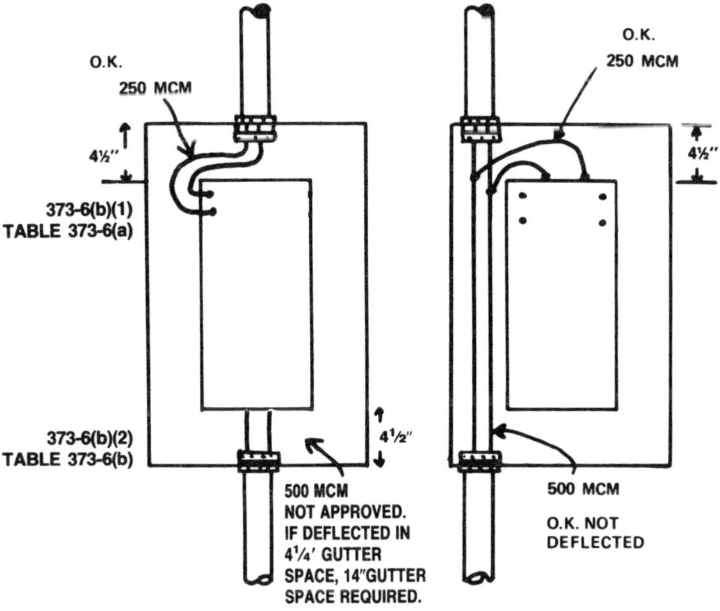

CONDUCTORS CAN NOT BE DEFLECTED UNLESS REQUIRED GUTTER SPACE IS PROVIDED.

N.E.C. Section 373-6 (a)

Wire Bending Space At Terminals

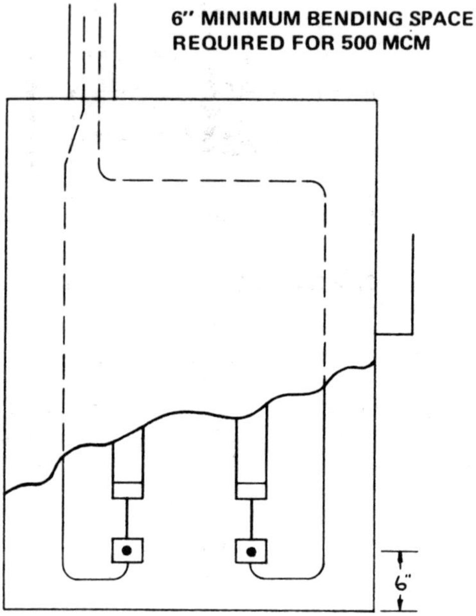

BENDING SPACE AT TERMINALS REQUIRED TO COMPLY WITH TABLE 373-6(a)

N.E.C. Sections 373-6 (a) & 374-9 (d)

Deflection Of Conductors In Gutter

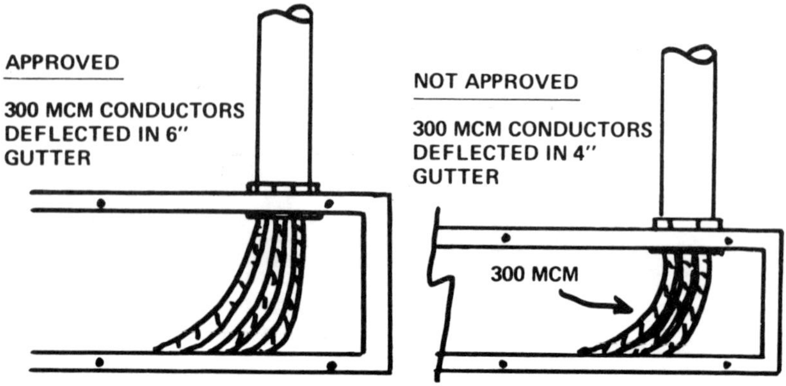

N.E.C. Section 373-6 (c)

Insulation At Bushings

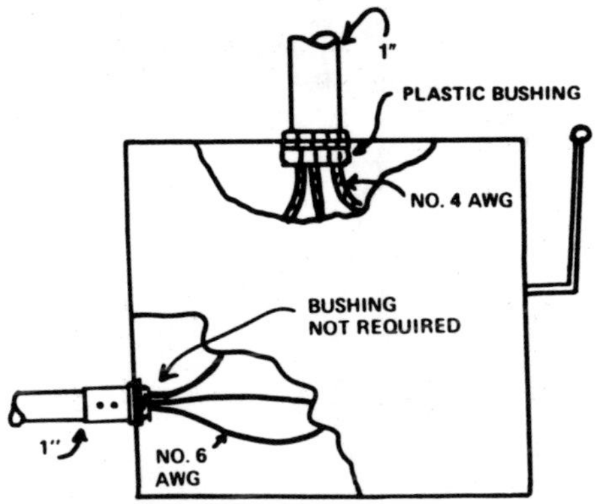

NO. 4 OR LARGER WIRES MUST HAVE INSULATING BUSHING OR INSULINER SLEEVE WITH BUSHING AT RACEWAY TERMINATIONS IN ENCLOSURES.

N.E.C. Section 373-8

Use Of Switches As Junction Boxes

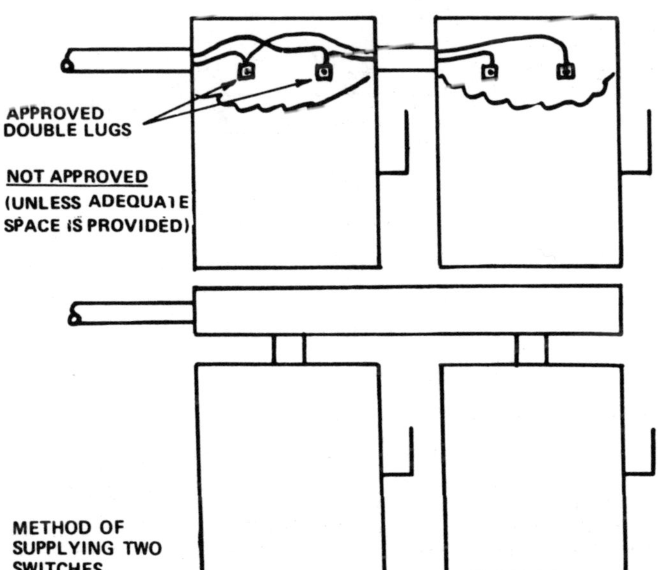

N.E.C. Sections 373-8, 430-10 & 430-8

Controller Case Not Approved As Pull Or Junction Box

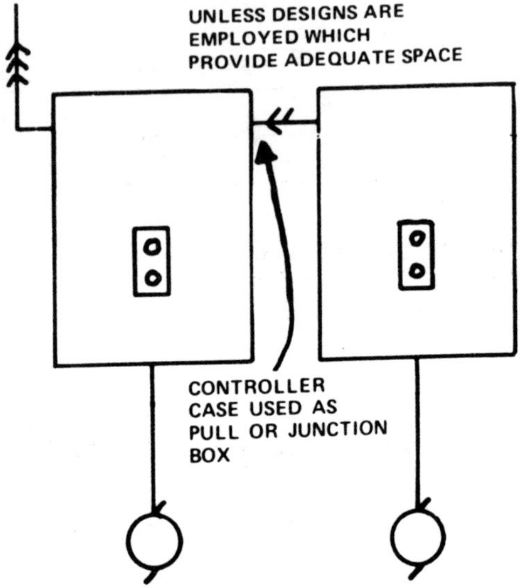

N.E.C. Section 374-5

Number Of Wires In Gutter

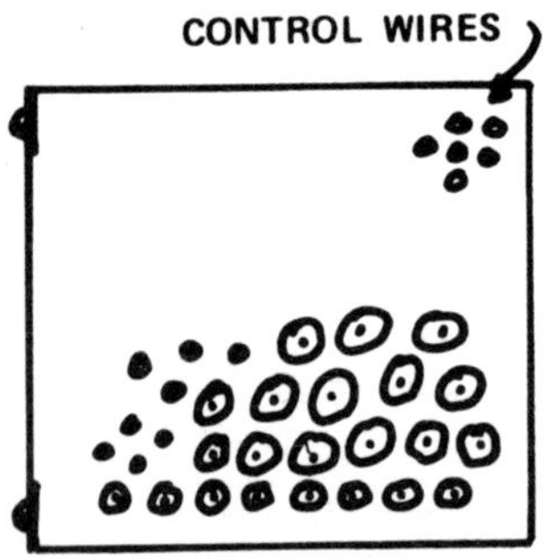

NEW GUTTER AND EXISTING

LIMITED TO 20% MAXIMUM FILL.

NOTE: WHERE MORE THAN 30 CURRENT CARRYING CONDUCTORS AT ANY CROSS SECTION AREA ARE INSTALLED THE CORRECTION FACTORS SPECIFIED IN NOTE 8 TO TABLES 310-16 THROUGH 310-19 SHALL BE APPLIED.

N.E.C. Section 374-6

Bare Bus Bars In Auxiliary Gutter

THE CONTINUOUS CURRENT RATING IN BARE COPPER BUS SHALL NOT EXCEED 1000 AMPERES PER SQUARE INCH OF CROSS SECTION AREA OR 700 AMPERES FOR ALUMINUM BUS.

N.E.C. Sections 380-3 & 373-8

Uses Of Switches As Junction Boxes

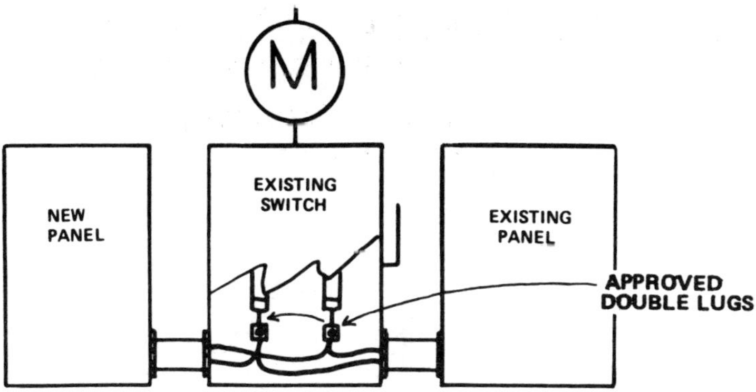

SWITCHES SUPPLYING CONDUCTORS MAY HAVE ADDITIONAL SETS TAPPED OFF IF ADEQUATE SPACE IS PROVIDED. THE BENDING SPACE FOR CONDUCTORS AT CIRCUIT BREAKERS AND SWITCHES IS REQUIRED TO COMPLY WITH 373-6.

NOTE: THIS DOES NOT VOID CODE SECTIONS DEALING WITH OVERCURRENT PROTECTION OF WIRE, ETC.

N.E.C. Section 380-8 (b)

Accessibility And Grouping

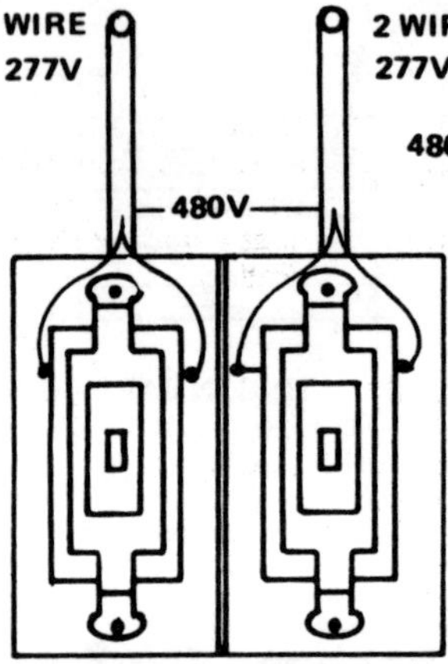

A PERMANENTLY INSTALLED BARRIER SHALL BE INSTALLED BETWEEN SWITCHES WITHIN SAME OUTLET BOX WHEN THE VOLTAGE BETWEEN ADJACENT SWITCHES *EXCEEDS 300 VOLTS.*

N.E.C. Section 380-13 (b)

Switches

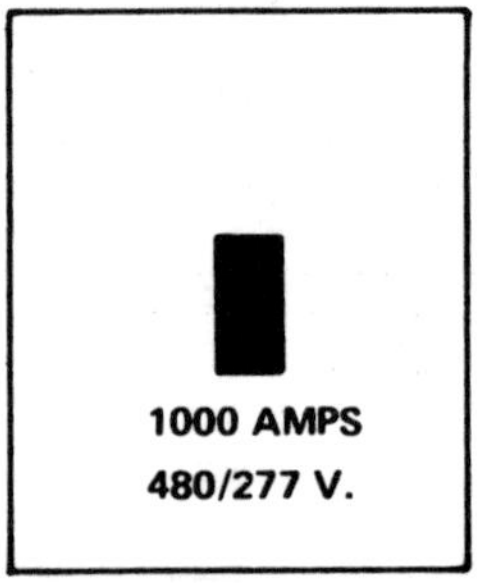

TO INTERRUPT CURRENTS GREATER THAN 1200 AMPERES AT 250 VOLTS OR LESS, OR 600 AMPERES AT 251 VOLTS TO 600 VOLTS, A CIRCUIT BREAKER OR SWITCH OF SPECIAL DESIGN IDENTIFIED FOR SUCH PURPOSE SHALL BE USED.

N.E.C. Section 380-14 (a)

A.C. General Use Snap Switch For Motor Loads

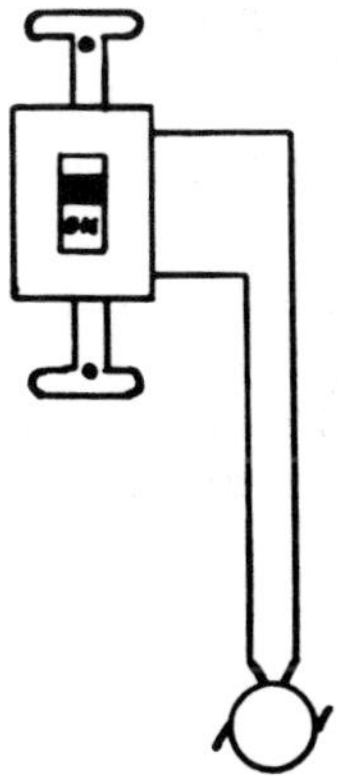

10A. 120V. A.C. ONLY General Use Snap Switch O.K. U.L. Considers these switches H.P. rated for loads up to 80% of Switch Rating.

1/3 HP 120V.
7.2A.

APPROVED

SWITCH FOR TUNGSTEN FILAMENT LAMPS MUST BE APPROVED FOR PURPOSE. MUST HAVE "T" RATING .

NOTE: U.L. CONSIDERS "A.C. GENERAL USE SNAP SWITCHES" O.K. FOR TUNGSTEN LOADS.

N.E.C. Section 380-14 (a), (b)

"T" Rated Snap Switches

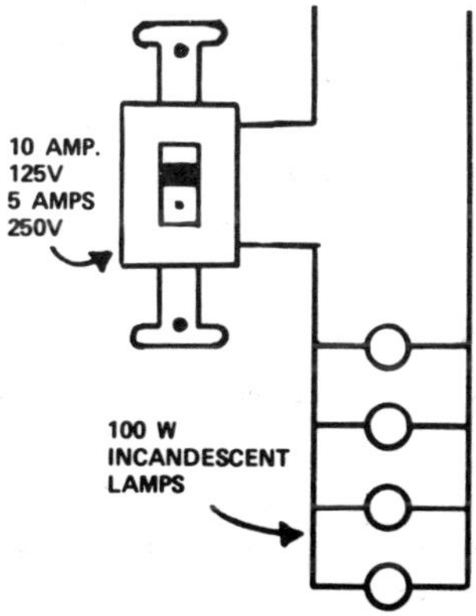

N.E.C. Section 380-14 (a) (1) (b) (2)

Switches For Inductive Loads

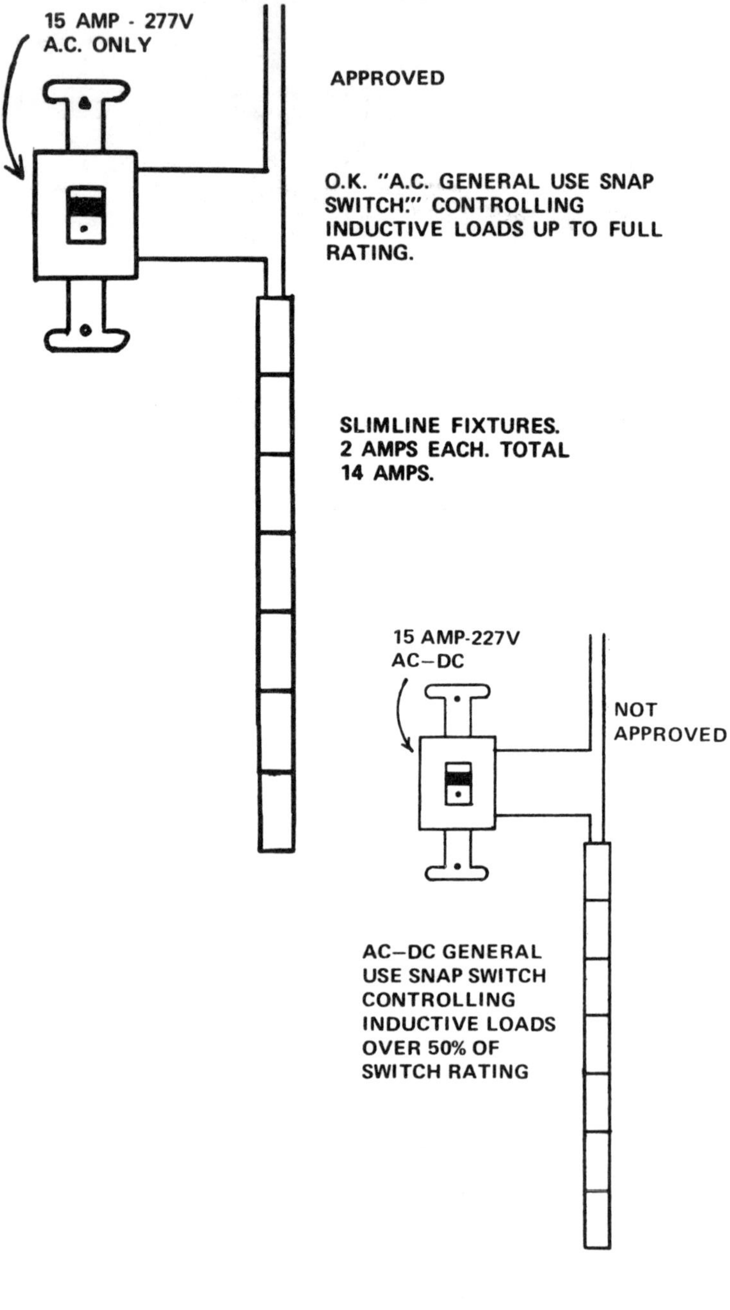

N.E.C. Section 384-10

Raceways Entering Switchboard

RACEWAYS
ENTERING OPEN BOTTOM SWITCHBOARD

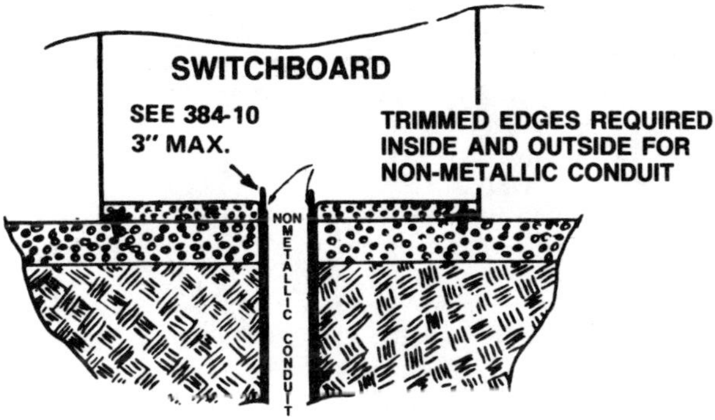

N.E.C. Section 384-10

Clearance From Busbars

SPACING BETWEEN BOTTOM
OF ENCLOSURE AND BUSBARS
IN
BUS ENCLOSURES

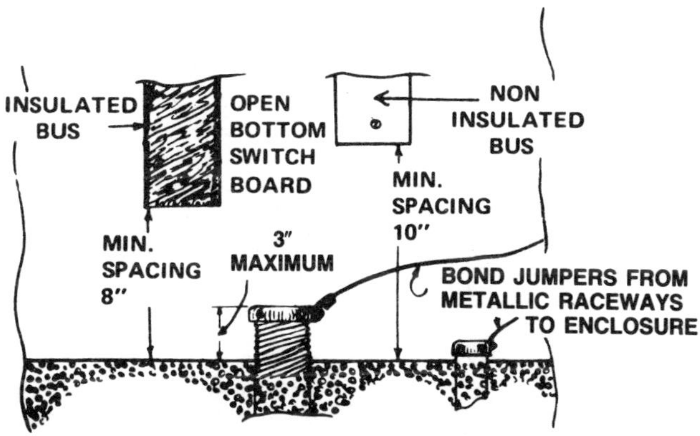

N.E.C. Section 384-13

Minimum Panelboard Rating

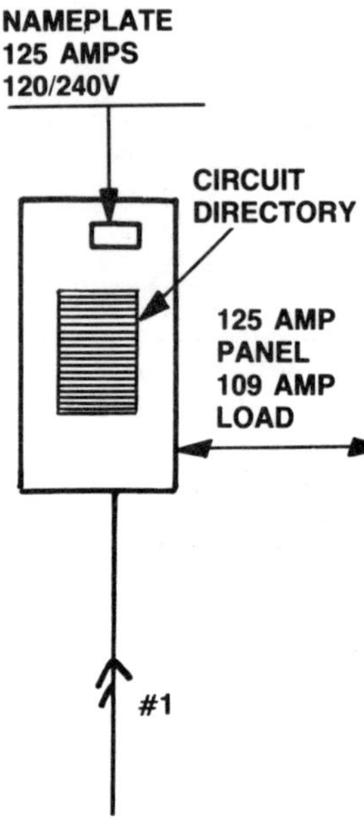

CIR	TYPE	WIRE	OUT LETS	LOAD Ø A	LOAD Ø B
A	LIGHT	12	10	15	
B	LIGHT	12	10		15
C D	SUB-1 MOTOR	6		45	45
E F	WATER HTR.	8	1	31	31
G	REC.	12	12	18	
H	REC.	12	12		18
			TOTAL AMPS	109	109

PANELBOARDS TO HAVE A RATING NOT LESS THAN THE FEEDER LOAD.

N.E.C. Section 384-16 (c)

Panelboard Devices For Continuous Loads

OVERCURRENT DEVICES IN PANELBOARDS AND USED TO SUPPLY CONTINUOUS LOADS SHALL HAVE A RATING OF NOT LESS THAN 125% OF THE CIRCUIT LOADING.

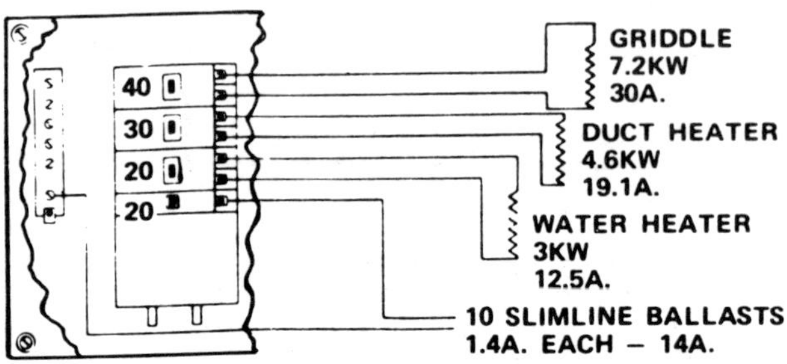

N.E.C. Sections 384-14, 16

Overcurrent Protection For Panelboards

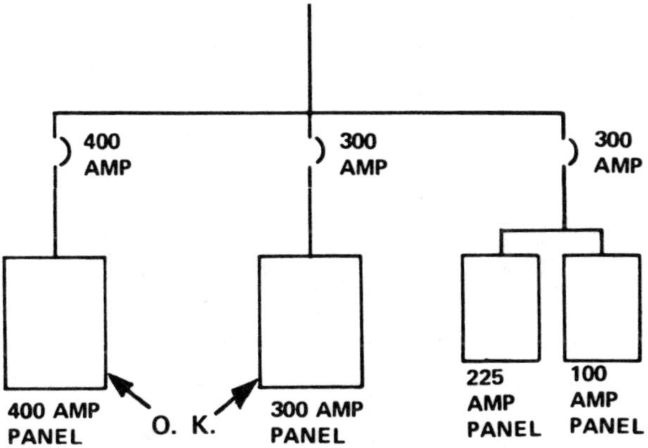

LIGHTING AND APPLIANCE BRANCH CIRCUIT PROTECTED NOT TO EXCEED THEIR RATING

NOT APPROVED
LIGHTING AND APP BRANCH CIR. PANELS PROTECTED AT OVER THEIR RATING.

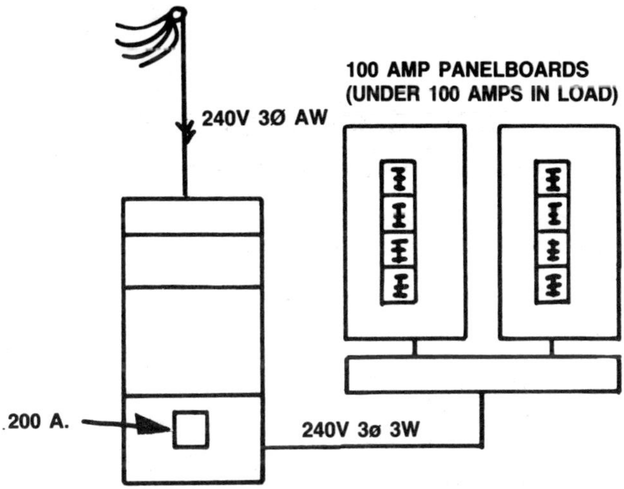

APPROVED: 100 AMP PANELBOARDS PROTECTED AT OVER THEIR RATINGS (NOT LIGHTING OR APPLIANCE BRANCH CIRCUIT PANELBOARDS, DUE TO, NO NEUTRAL IN SYSTEM)

N.E.C. Section 400-4

Voltage Rating Of Cord

N.E.C. Sections 400-8 & 400-9

Unapproved Use Of Cords

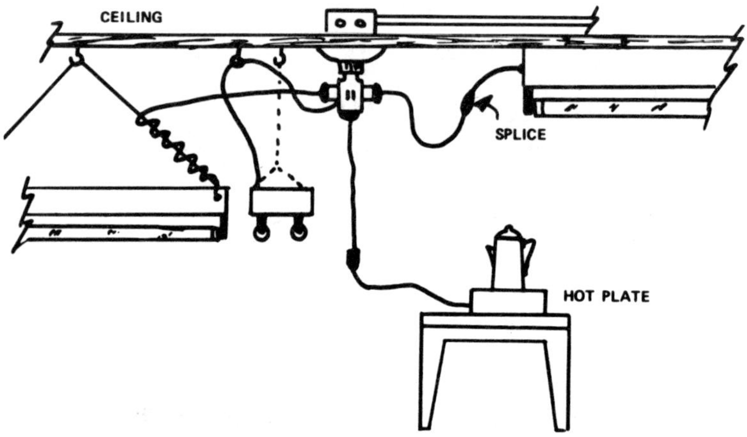

N.E.C. Section 410-11

Temperature Limits Of Wires In Outlet Boxes

OPERATING TEMPERATURE
SHALL NOT EXCEED THE
TEMPERATURE RATING
OF THE CONDUCTORS

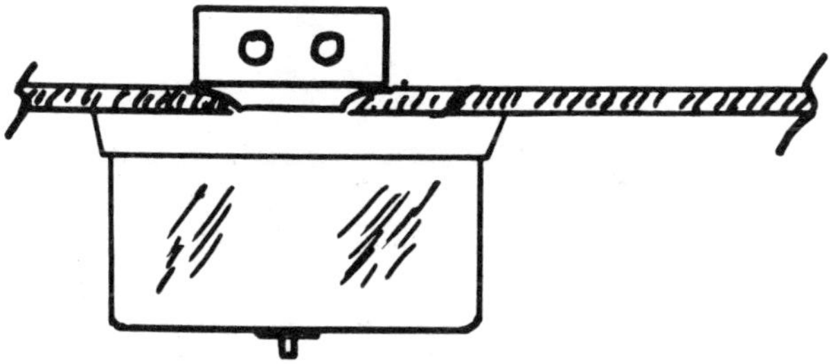

N.E.C. Section 410-11

Conductors In Outlet Boxes

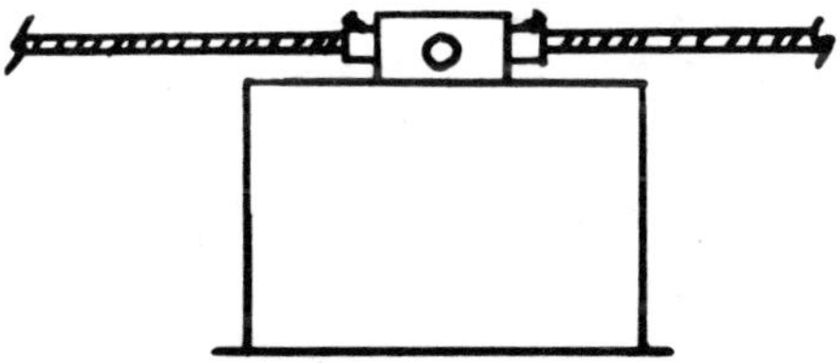

<u>NOT APPROVED</u> (MORE THAN TWO WIRES IN BOX ON INCANDESCENT FIXTURE, UNLESS LABELED OTHERWISE)

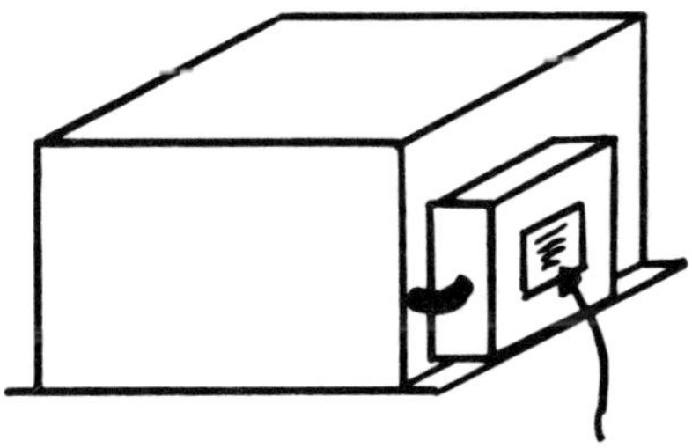

LABEL

<u>APPROVED</u>
THE BOX MAY BE USED
AS PART OF THE WIRING
SYSTEM PER CONDITIONS
STATED ON LABEL

XYZ LIGHT CO.
THIS BOX <u>APPROVED</u>
FOR UP TO 12 #12
CIRCUIT WIRES

N.E.C. Sections 410-15, 16

Fixture Supports Require Listing

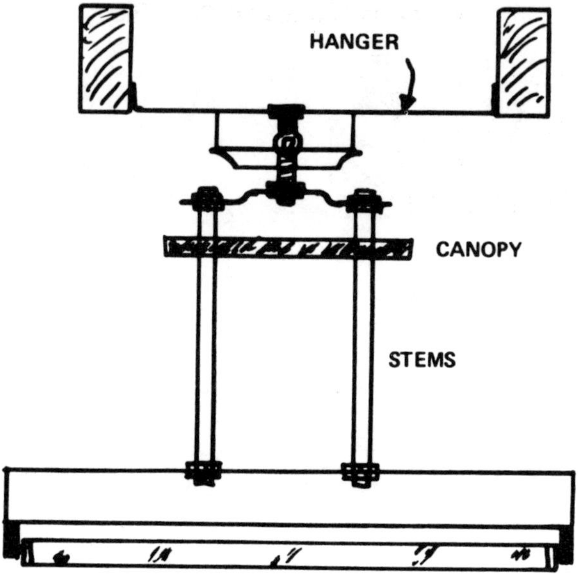

BOX HANGERS, FIXTURE STUDS AND HICKEYS, HANGER STRAPS, RODS AND STEMS, AND THEIR MEANS OF FASTENING SHALL BE OF THE APPROVED TYPE.

FIXTURES SUPPORTED BY A JUNCTION BOX MAY NOT EXCEED 50 LBS.

ALSO, WHERE CLIPS DESIGNED TO RETAIN THE FIXTURES IN PLACE ARE USED THE CLIP SHALL BE IDENTIFIED AS ONE SUITABLE FOR THE FIXTURE AND TYPE OF FRAMING MEMBER.

410-20. *EQUIPMENT GROUNDING CONDUCTOR ATTACHMENT.*

FIXTURES WITH EXPOSED METAL PARTS SHALL BE PROVIDED WITH A MEANS FOR CONNECTING AN EQUIPMENT GROUNDING CONDUCTOR FOR SUCH FIXTURES.

N.E.C. Section 410-28 (d)

Limitations Of Splices In Outlet Box

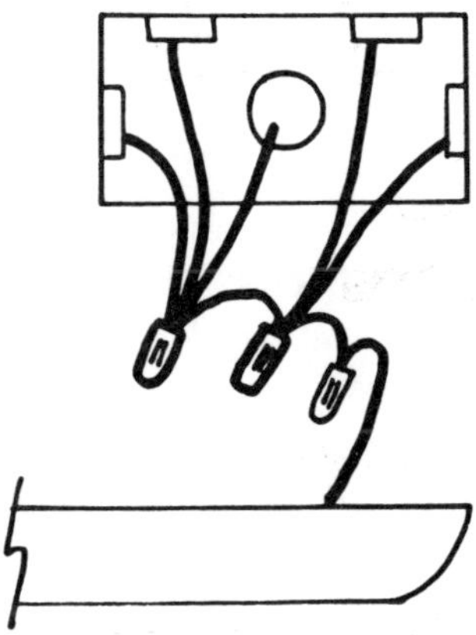

NOT APPROVED. MORE THAN ONE JOINT IN ONE SET OF WIRES IN OUTLET BOX.

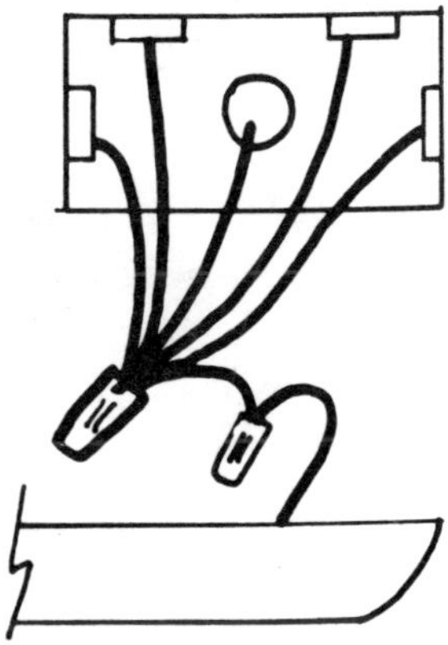

O.K. ONE JOINT PLUS FIXTURE WIRE JOINT. O.K. IN OUTLET BOX.

N.E.C. Section 410-30 (c)

Cord Connected Electric Discharge Fixtures

APPROVED

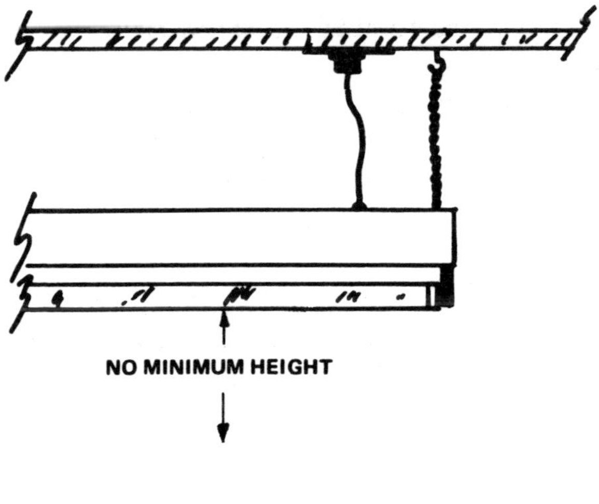

N.E.C. Section 410-31, Ex. 2

End To End Connected Fixture

CONDUCTORS SUPPLYING END-TO-END CONNECTED FIXTURES SHALL HAVE INSULATION SUITABLE FOR 90° WITHIN 3" OF A BALLAST.
NOTE: THW WITHIN ELECTRIC DISCHARGE LIGHTING EQUIPMENT (1000 OPEN CIRCUIT VOLTS OR LESS) MAY BE USED. (SEE TABLE 310-13 FOR THW.)

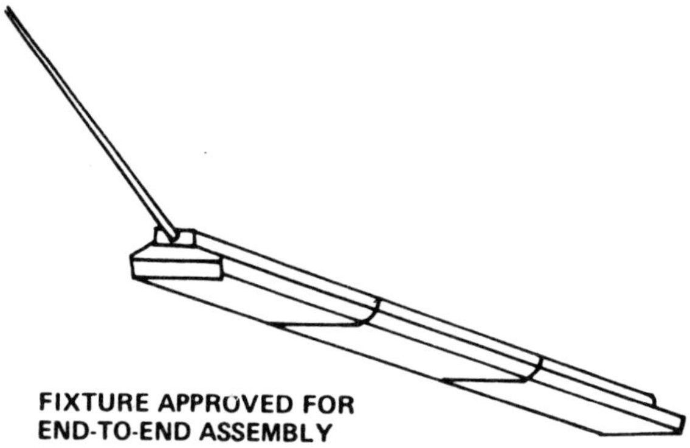

N.E.C. Section 410-31, Ex. 2 & 3

Multiwire Branch Circuit

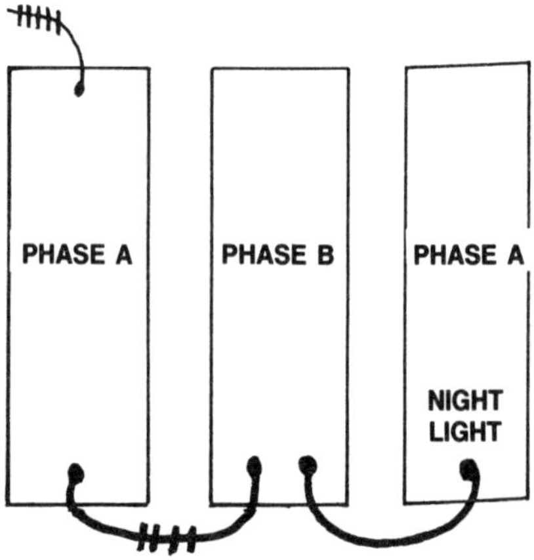

A MULTI WIRE BRANCH CIRCUIT AND AN ADDITIONAL 2 WIRE CIRCUIT IS PERMITTED TO SUPPLY END TO END CONNECTED FIXTURES.

N.E.C. Section 410-54 (b)

Switching

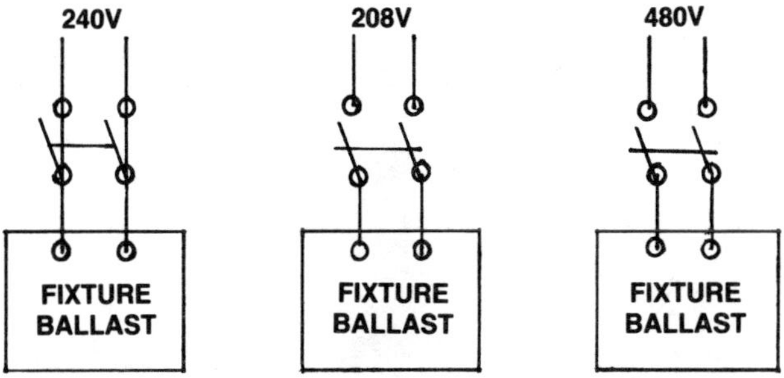

BOTH "HOT" CONDUCTORS OF A 2-WIRE UNGROUNDED CIRCUIT TO ELECTRIC DISCHARGE-LAMP AUXILIARY EQUIPMENT MUST SIMULTANEOUSLY BE DISCONNECTED.

N.E.C. Sections 410-57, 58

Receptacles In Damp Or Wet Locations

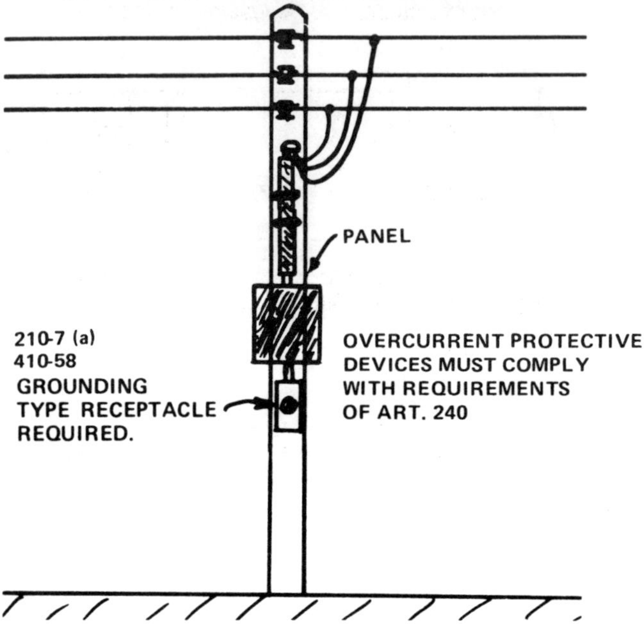

N.E.C. Section 410-64

Fixtures Required To Be Listed For Recessing

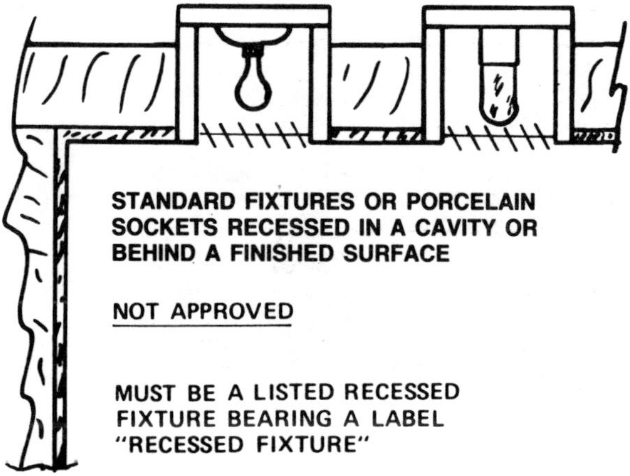

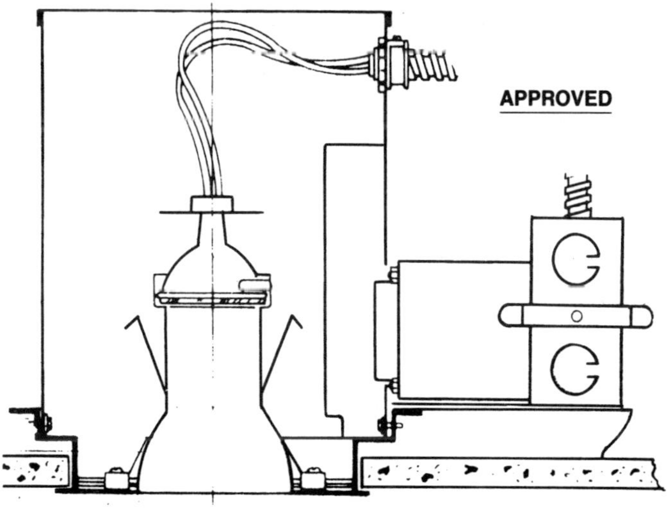

N.E.C. Section 410-65 (c)

Recessed Incandescent Fixtures

(c) *RECESSED INCANDESCENT FIXTURES.* INCANDESCENT FIXTURES SHALL HAVE THERMAL PROTECTION AND SHALL SO BE IDENTIFIED AS THERMALLY PROTECTED.

EXCEPTION NO. 1: RECESSED INCANDESCENT FIXTURES IDENTIFIED FOR USE AND INSTALLED IN POURED CONCRETE.

EXCEPTION NO. 2: LISTED RECESSED INCANDESCENT FIXTURES THAT PROVIDE, BY CONSTRUCTION DESIGN, THE EQUIVALENT TEMPERATURE PERFORMANCE CHARACTERISTICS OF THERMALLY PROTECTED FIXTURES AND ARE SO IDENTIFIED.

N.E.C. Section 410-66 (b)

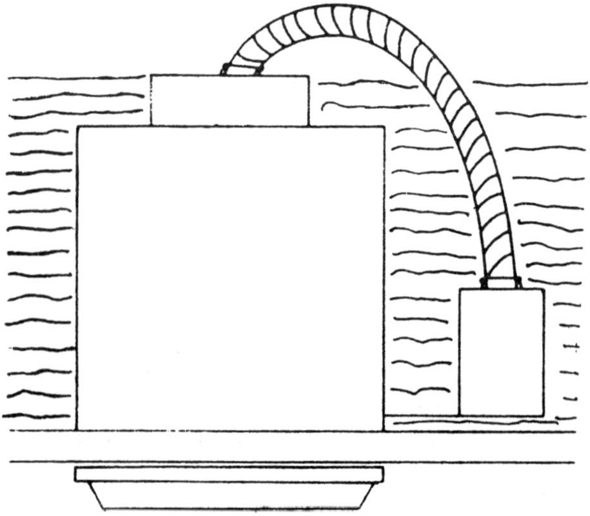

ABNORMAL CONDITIONS UNLESS MARKED FOR SUCH USE.

N.E.C. Section 410-66

Clearance Of Recessed Fixtures

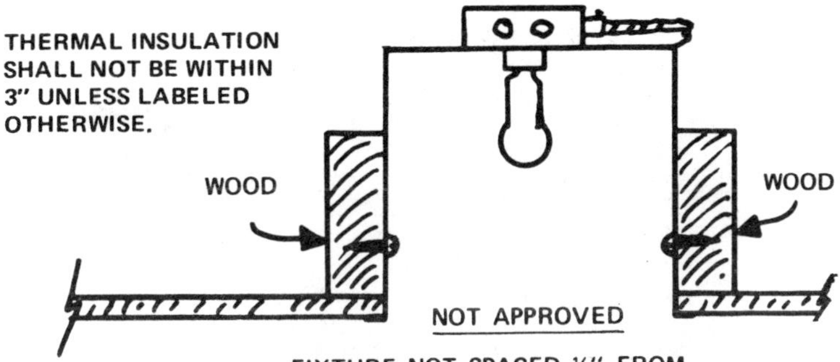

FIXTURE NOT SPACED ½" FROM WOOD. (MANUFACTURER PROVIDES APPROVED MEANS OF SUPPORT)

N.E.C. Section 410-76 (b)

1½" Spacing Required

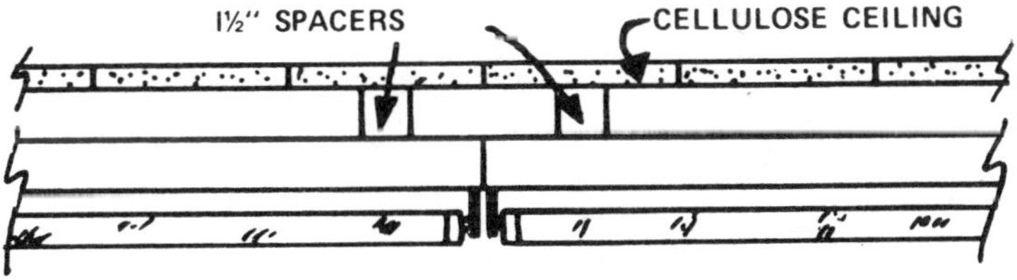

FIXTURES CONTAINING BALLASTS REQUIRE 1½" SPACING FROM CELLULOSE SURFACES OR BE APPROVED FOR THE PURPOSE.

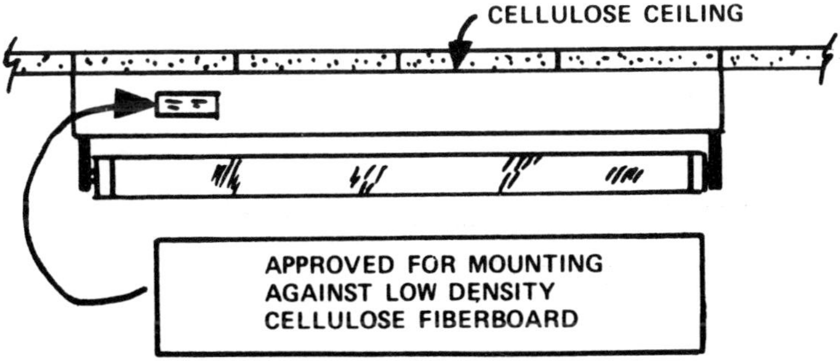

LOOK FOR LABEL

N.E.C. Section 422-22 (d)

Attachment Plugs Operating At Over 300 Volts

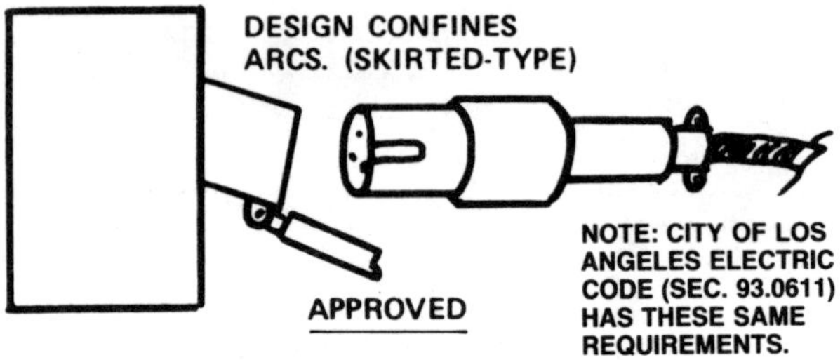

N.E.C. Section 422-28 (e)

Single Appliances

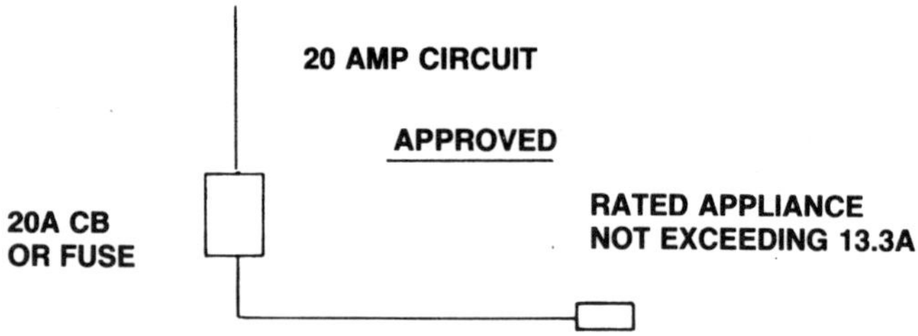

WHEN RATED OVER 13.3A: OVERCURRENT PROTECTION *SHALL NOT EXCEED 150 PERCENT* OF THE APPLIANCE RATING, WHERE THE CIRCUIT SUPPLIES ONLY A SINGLE NON MOTOR-OPERATED APPLIANCE.

N.E.C. Section 424-3 (a) & (b)

Fixed Space Heater Circuits

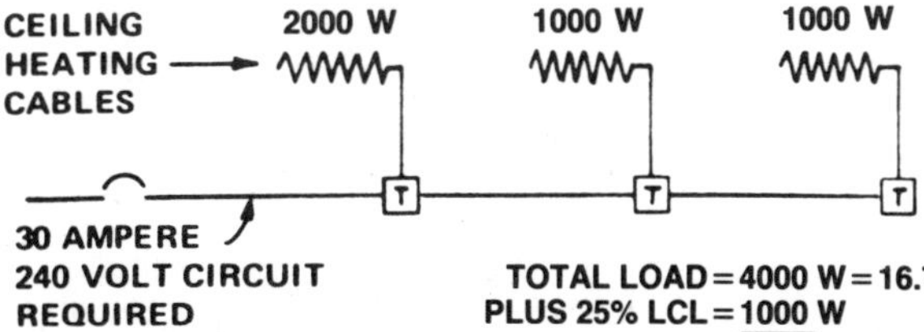

TOTAL LOAD = 4000 W = 16.7 AMPS
PLUS 25% LCL = 1000 W
5000 W = 20.8 AMPS

TWO OR MORE *FIXED SPACE HEATERS* ARE PERMITTED ON A 15, 20, OR 30 AMPERE CIRCUIT – PROVIDED THE TOTAL LOAD DOES NOT EXCEED 80% OF THE CIRCUIT.

TYPICAL SPACE HTRS INCLUDE:

DUCT HTRS	BASEBOARD HTRS
PANELS	WALL INSERT UNITS
CABLES (FLOOR OR CEIL)	ETC.

THE RATING OR SETTING OF OVERCURRENT PROTECTIVE DEVICES SHALL BE PERMITTED IN ACCORDANCE WITH SECTION 240-3, EXCEPTION NO. 4.

N.E.C. Section 424-13

Spacing From Combustible Material

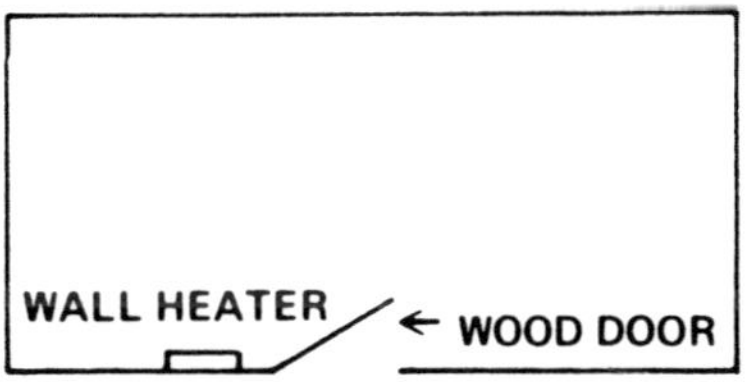

NOT APPROVED – DOOR COULD SWING IN FRONT OF HEATER.

FIXED ELECTRICAL SPACE HEATING EQUIPMENT SHALL BE INSTALLED TO PROVIDE SPACING BETWEEN THE EQUIPMENT AND COMBUSTIBLE MATERIAL.

N.E.C. Section 424-19

Motor-Driven Heating Equipment

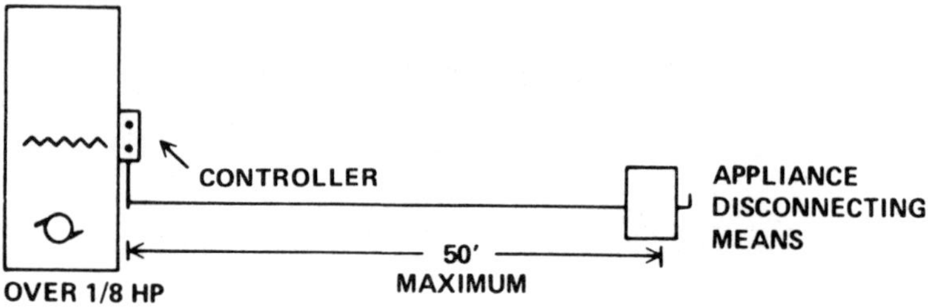

A SWITCH OR CIRCUIT BREAKER MAY SERVE AS THE DISCONNECTING MEANS FOR A MOTOR-DRIVEN HEATING APPLIANCE WHEREIN SIGHT OF THE MOTOR CONTROLLER AND NOT OVER 50' MAXIMUM [430-102(a)]

N.E.C. Section 424-19

Appliance Disconnecting Means

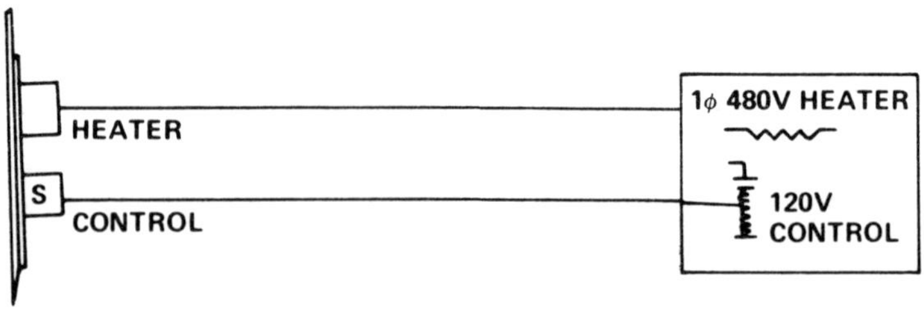

THE DISCONNECTING MEANS SUPPLYING HEATING EQUIPMENT FROM MORE THAN 1 SOURCE *SHALL BE GROUPED* AND *IDENTIFIED.*

N.E.C. Section 424-22 (b)

Overcurrent Protection

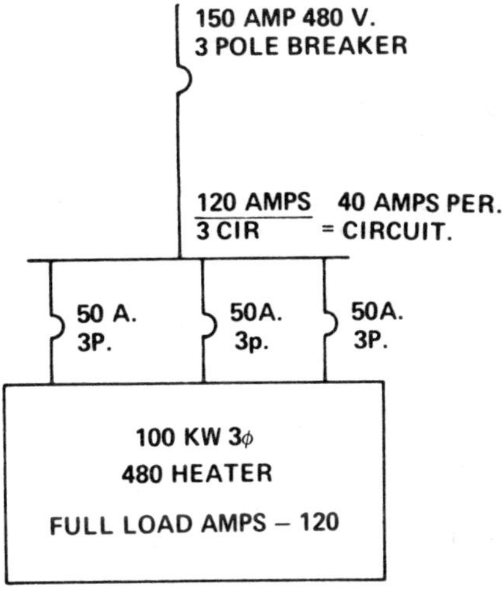

ELECTRIC SPACE HEATING EQUIPMENT EMPLOYING RESISTANCE-TYPE HEATING ELEMENTS RATED *MORE THAN 48 AMPERES* SHALL HAVE HEATING ELEMENTS *SUBDIVIDED.* EACH SUBDIVIDED LOAD SHALL NOT EXCEED 48 AMPERES AND SHALL BE PROTECTED AT NOT MORE THAN 60 AMPERES.

NOTE: FACTORY ASSEMBLED OVERCURRENT PROTECTIVE DEVICES WITHIN THE HEATER ENCLOSURE NEED NOT BE READILY ACCESSIBLE.

N.E.C. Section 424-22 (b)

Multi-stage Duct Heaters

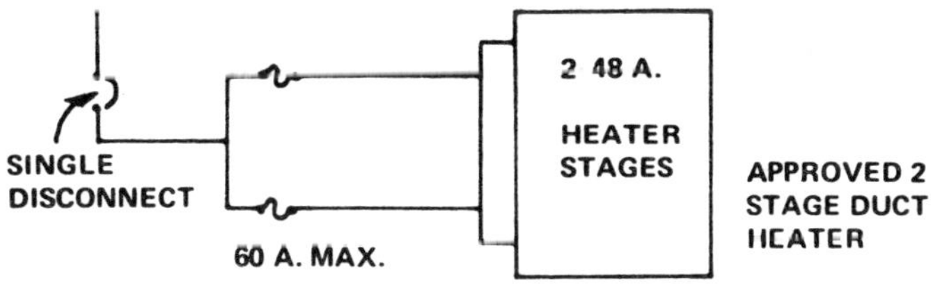

HEATING ELEMENTS OF A DUCT HEATER MORE THAN 48 AMPS SHALL BE SUBDIVIDED IN SECTIONS NOT LARGER THAN 48 AMPS EACH.

OVERCURRENT PROTECTION FOR EACH STAGE SHALL NOT BE LARGER THAN 60 AMPS IN RATING.

WHERE A SUBDIVIDED LOAD IS LESS THAN 48 AMPERES THE RATING OF THE OVERCURRENT PROTECTIVE DEVICE SHALL COMPLY WITH SECTION 424-3 (b).

EXCEPTION: AS PROVIDED IN SECTION 424-72 (a).

N.E.C. Section 424-58

Duct Heaters

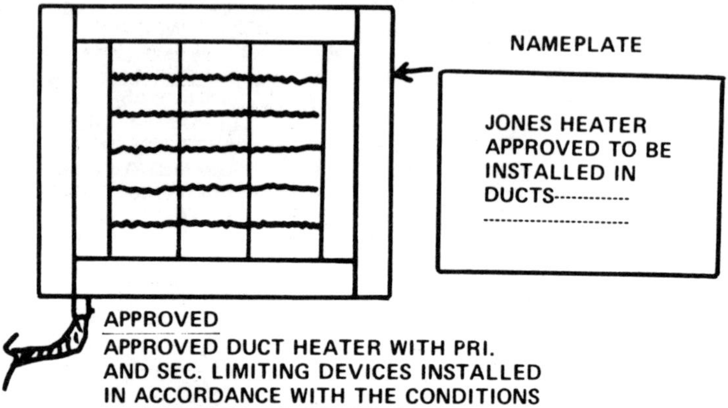

APPROVED

APPROVED DUCT HEATER WITH PRI. AND SEC. LIMITING DEVICES INSTALLED IN ACCORDANCE WITH THE CONDITIONS OF APPROVAL AS STATED ON THE DUCT HEATER.

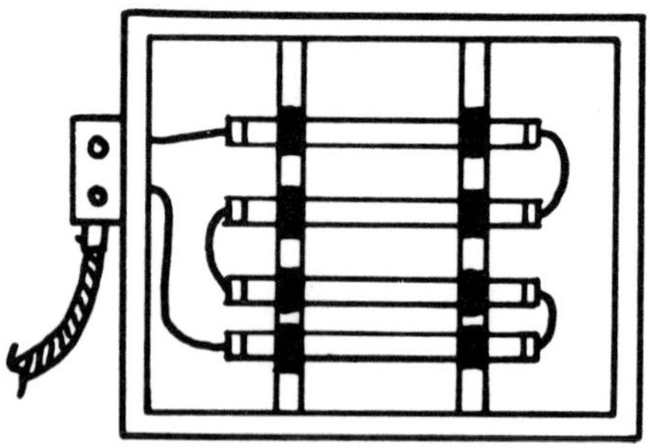

NOT APPROVED

FIELD ASSEMBLED HEATING ELEMENTS USED AS AIR HEATER IN DUCT

N.E.C. Sections 424-63 & 424-64

Installation Of Duct Heater

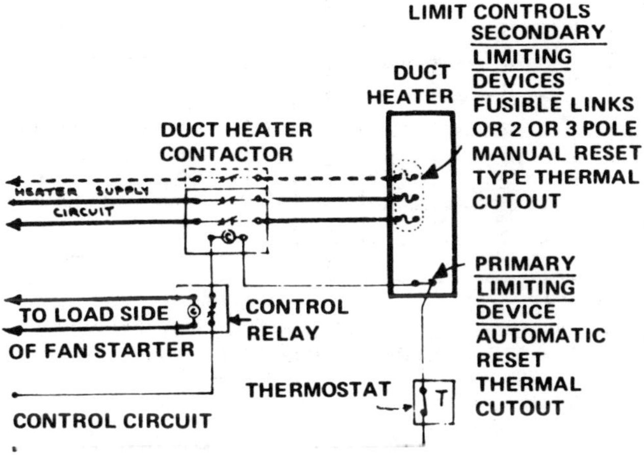

DUCT HEATER CONTACTOR MUST BE APPROVED FOR THE DUCT HEATER IT CONTROLS.

BOTH PRIMARY AND SECONDARY LIMITING DEVICES MUST BE FACTORY INSTALLED AND APPROVED UNITS.

N.E.C. Section 430-6

Name Plate Rating Used To Determine Various Ampacities

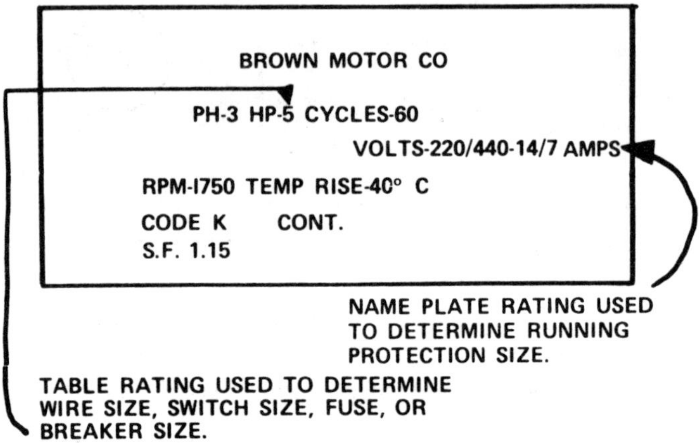

N.E.C. Section 430-23 (c)

Secondary Conductors For Wound Rotor Motors

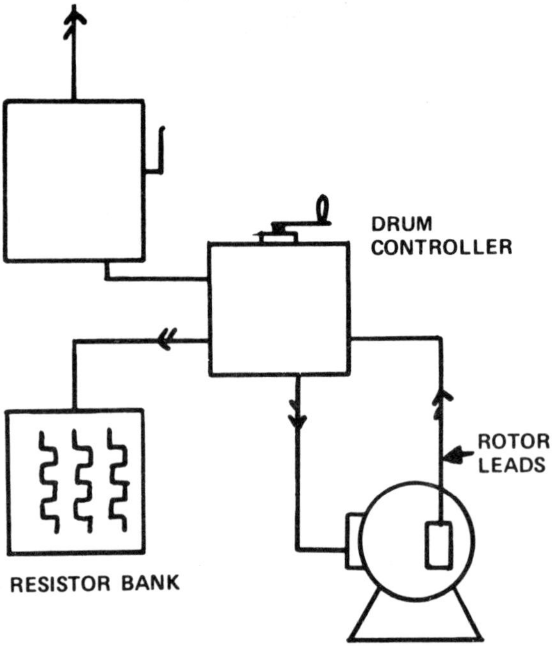

WIRES BETWEEN CONTROLLER AND RESISTORS, SIZE DETERMINED FROM TABLE 430–23(c) DEPENDING ON STARTING DUTY, ETC)

N.E.C. Section 430-32

Types Of Running Protection

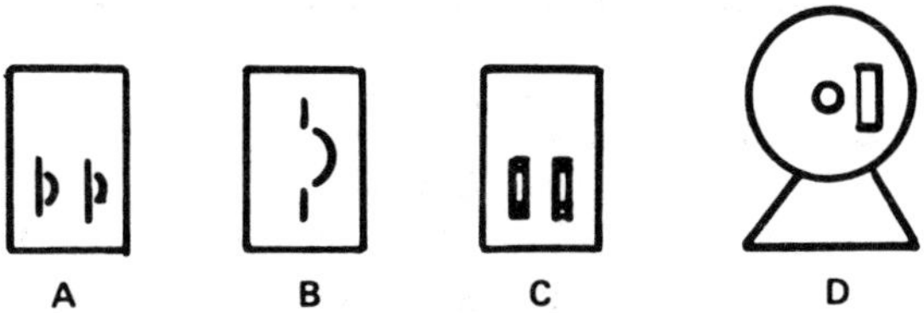

A – RELAYS THAT ARE A PART OF AN APPROVED CONTROLLER.

B – CIRCUIT BREAKER OF PROPER SIZE AND TIME DELAY.

C – FUSES OF PROPER SIZE AND TIME DELAY.

D – PROTECTIVE DEVICE INTEGRAL WITH MOTORS.

N.E.C. Sections 430-32 & 430-34

Size Of Motor Running Protection

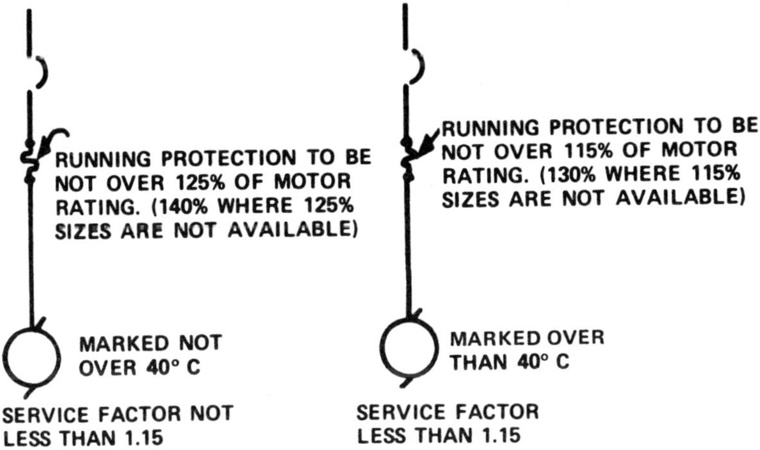

N.E.C. Section 430-32 (a) 3 & (c) 3

U.L. Listed, Factory-wired Machine Considered As Having Running Protection

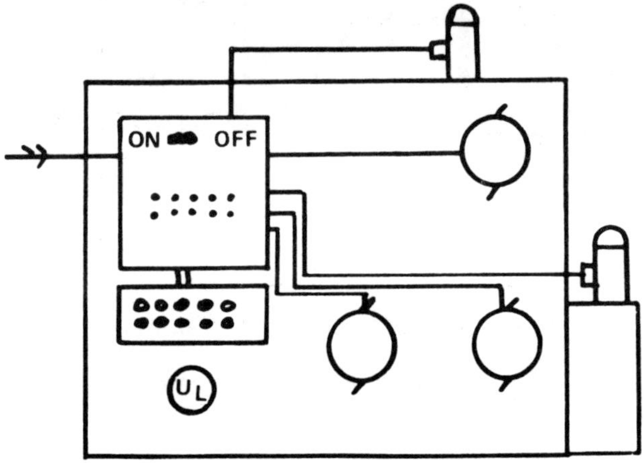

U.L. FACTORY WIRED CLEANING MACHINE. MOTORS ALL CONSIDERED AS HAVING RUNNING PROTECTION.

N.E.C. Sections 430-32 (b)

Motor Over 50 Feet From Starter. R.P. Required

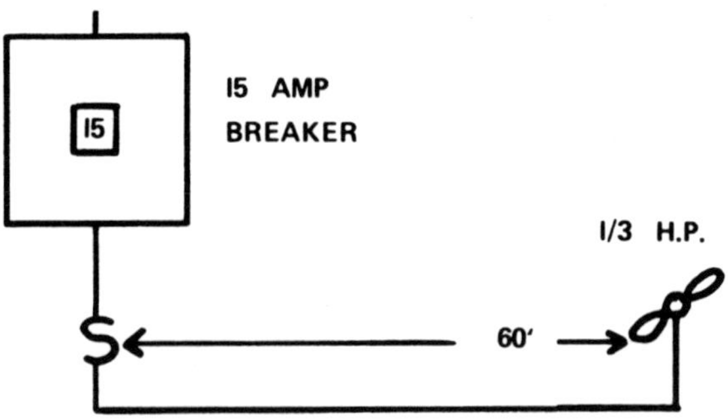

NOT APPROVED

NO RUNNING PROTECTION, MOTOR MORE THAN 50′ FROM CONTROLLER.

N.E.C. Section 430-32 (b)

Motor Not Within Sight Of Starter. Running Protection Required

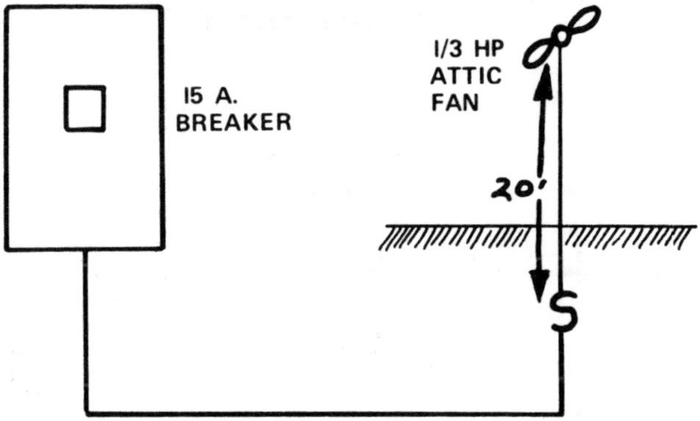

NOT APPROVED

NO RUNNING PROTECTION, MOTOR NOT IN SIGHT OF CONTROLLER.

N.E.C. Section 430-32 (c)

Motor Automatically Started Running Protection Required

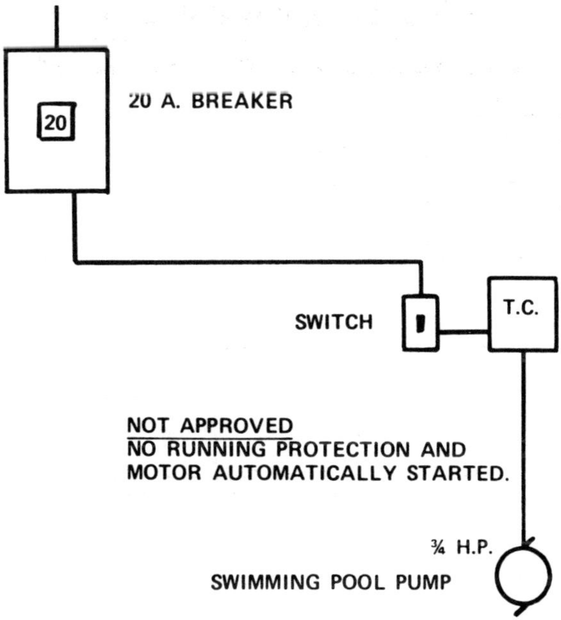

NOT APPROVED

NO RUNNING PROTECTION AND MOTOR AUTOMATICALLY STARTED.

N.E.C. Section 430-32 (c)

Motor Automatically Started. R.P. Required

RUNNING PROTECTION IS <u>NOT REQUIRED</u> FOR MOTORS OF ONE H.P. OR LESS ON 15 AMP OR 20 AMP 120 VOLT, OR ON 15 AMP 240 OR 480 V. CIRCUITS.

WHERE:

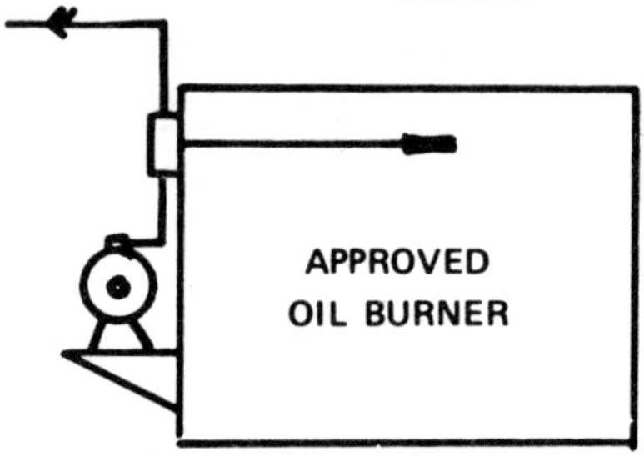

APPLIES TO CLOCK MOTORS, ETC.

MOTOR IS ON AN APPROVED ASSEMBLY EQUIPPED WITH OTHER SAFETY CONTROLS, ETC.

MOTOR IS OF TYPE HAVING WINDINGS OF SUFFICIENT IMPEDANCE TO PREVENT OVERHEATING, ETC.

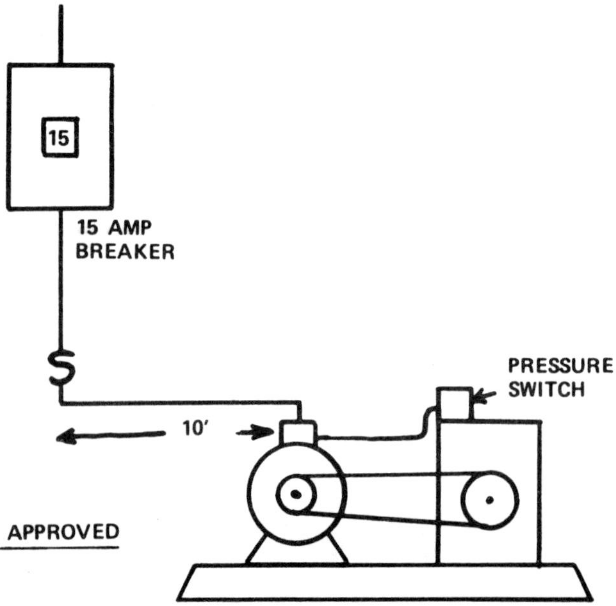

<u>NOT APPROVED</u>

NO RUNNING PROTECTION ON AN AUTOMATICALLY STARTED MOTOR.

AIR COMPRESSOR

N.E.C. Section 430-33

Short Time Duty Motors

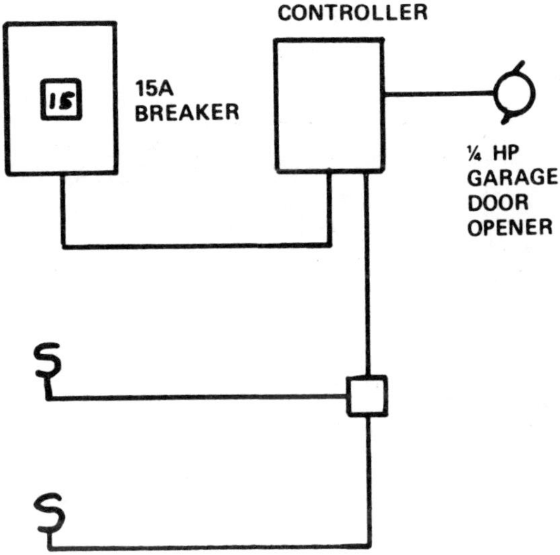

O.K. WITHOUT RUNNING PROTECTION,
AS DUTY IS SHORT TIME.

N.E.C. Sections 430-34, 430-40 & 430-55

Time Delay Required For Fuses Or Circuit Breakers Used As Running Protection

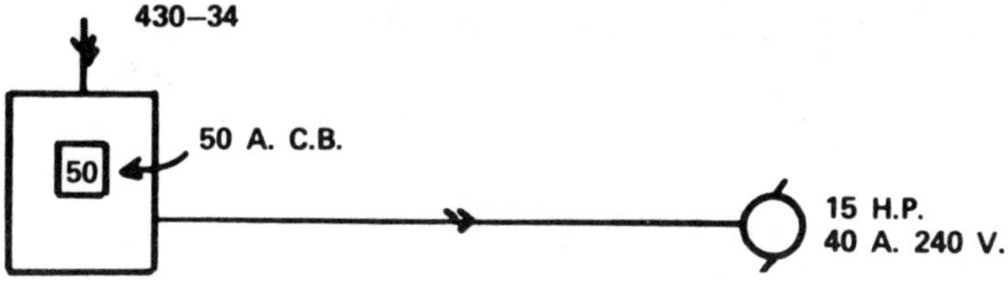

CIRCUIT BREAKER O.K. AS COMBINED RUNNING AND STARTING PROTECTION WHEN MEETING REQUIREMENTS OF 430-32 FOR TIME DELAY, ETC.

STANDARD FUSES NOT APPROVED AS COMBINED STARTING AND RUNNING PROTECTION (INSUFFICIENT TIME DELAY)

N.E.C. Section 430-37

Running Overcurrent Units

OVERLOAD RELAY IS REQUIRED IN EACH LEG OF 3Ø MOTOR

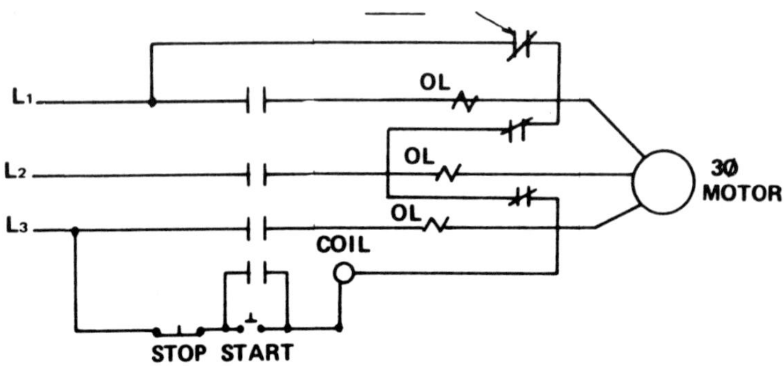

N.E.C. Sections 430-38 & 430-55

Circuits Breakers Used For Running Protection

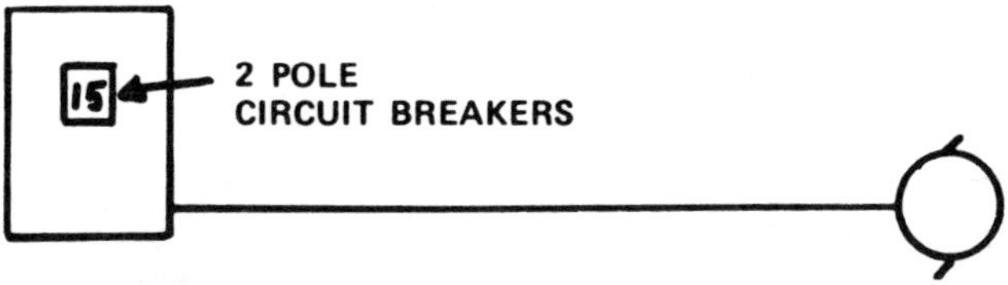

2 POLE C.B. MUST BE USED WHERE
CIRCUIT BREAKERS ARE USED AS
A DISCONNECTING MEANS.
OR 2 SINGLE POLE
CIRCUIT BREAKERS WITH
HANDLE BARS.

2 H.P. 12 A 240 V 1Ø

N.E.C. Section 430-42 (c)

Cord And Plug Connected Motor Operated Appliance

CORD & PLUG CONNECTED
MOTOR OPERATED APPLIANCE
LOCATION OF RUNNING
PROTECTION.

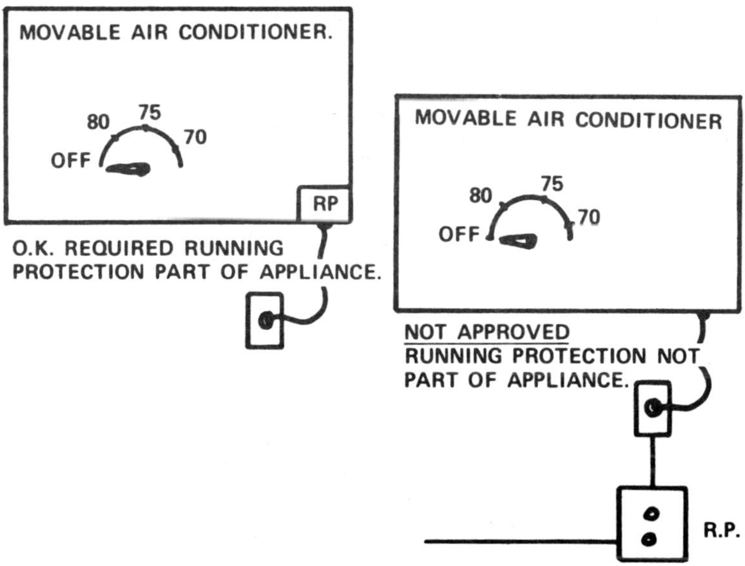

N.E.C. Section 430-53 (a) (1)

Small Motors On Same Circuit

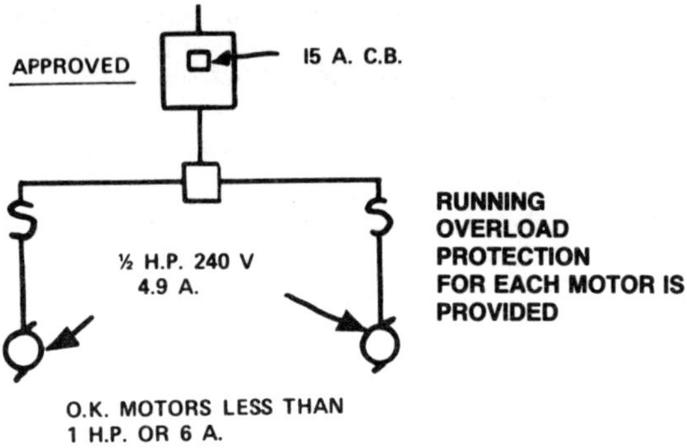

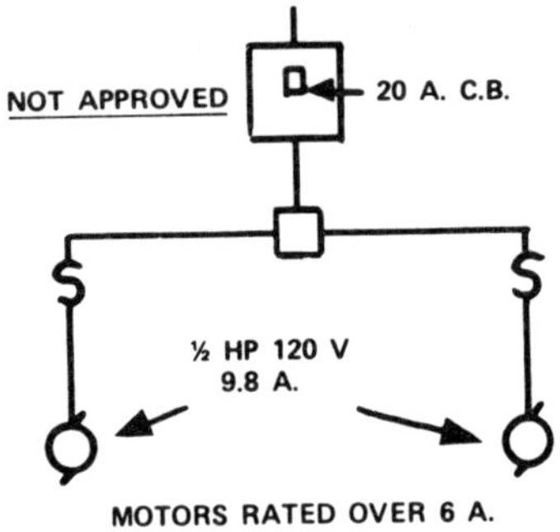

N.E.C. Section 430-53 (a)

Several Motors Or Loads On One Breaker Circuit

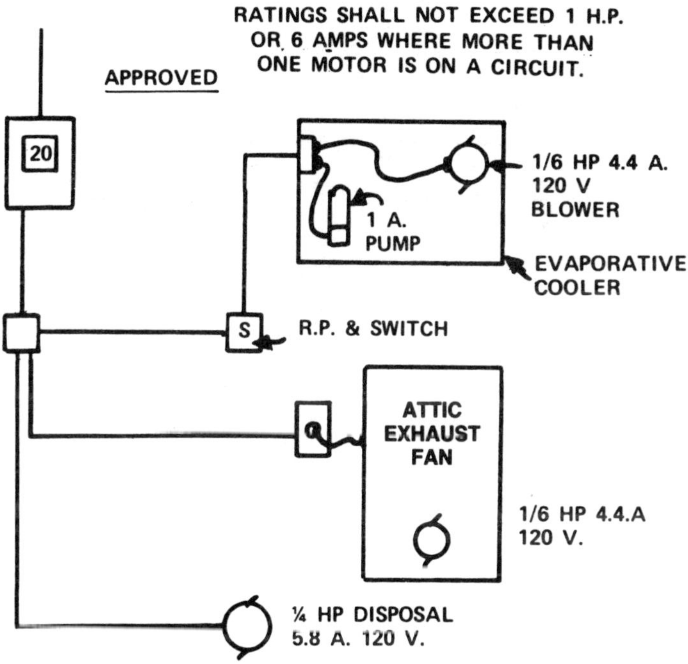

4.4+4.4+5.8+1 = 15.6 A.

GROUPED MOTORS O.K. ON CIRCUIT PROVIDED BREAKER HAS SUFFICIENT TIME DELAY TO START MOTORS.

N.E.C. Sections 430-53(a) & 430-112, Ex.(b)

Disconnect For A Group Of Motors

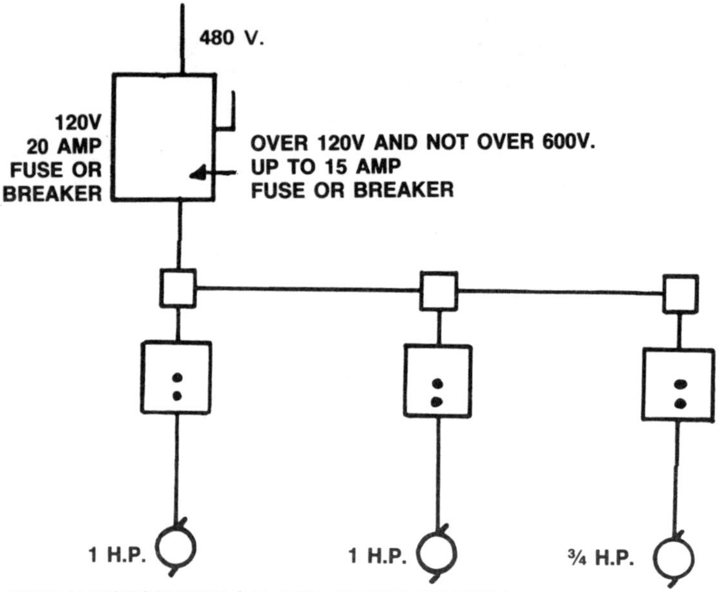

SINGLE DISCONNECT O.K. FOR GROUP OF MOTORS PERMITTED ON A SINGLE BRANCH CIRCUIT. (1 H.P. OR LESS AND 6 AMPS OR LESS)

N.E.C. Section 430-53 (c)

Multi-motored Machine Switch And Conductor Size

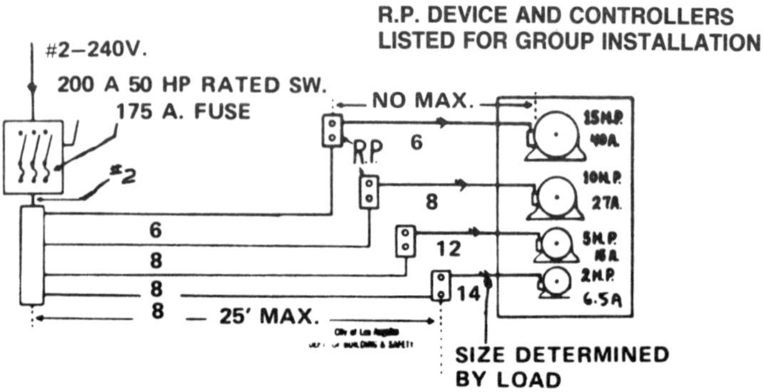

TAPS FOR MULTIMOTORED MACHINE CIRCUIT, (MUST BE MINIMUM OF 1/3 THE SIZE OF CIRCUIT CONDUCTOR.)

N.E.C. Section 430-57 & 152

Capacity Of Fuse Holders

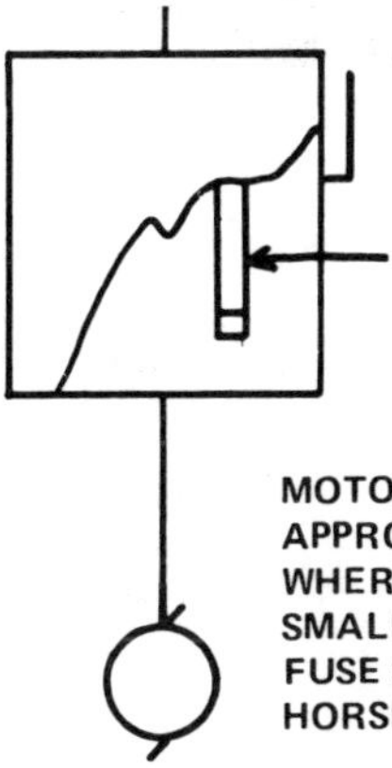

MOTOR BRANCH CIRCUIT FUSES HAVING APPROPRIATE TIME DELAY ARE REQUIRED, WHERE THE FUSE HOLDERS ARE OF A SMALLER SIZE THAN THE MAX. PERMISSIBLE FUSE SIZE. SEE SECTION 430-109 FOR HORSEPOWER RATING REQUIREMENTS.

ALL OF THE FOLLOWING LOADS OPERATE TOGETHER AND ARE CALCULATED AT 240 VOLTS 3 PHASE 60 HERTZ. ALL MOTORS ARE SQUIRREL CAGE, CODE LETTER B, FULL VOLTAGE START MOTORS.

20 H.P. MOTOR	**50.4 AMPS**
3 H.P. MOTOR	**9.6 AMPS**
1 H.P. MOTOR	**3.6 AMPS**
⅙ H.P. MOTOR	**0.7 AMPS**
15 KVA HEATER	**36.1 AMPS**

N.E.C. Section 430-57 & 62 (a)

Branch Circuit Fuse Size

EXAMPLE I
TABLE 430-152 LISTS THIS TYPE OF MOTOR FOR MAXIMUM NONTIME DELAY FUSE SETTING OF 250% OF FULL LOAD CURRENT.

20 HP. MOTOR	54.0 AMPS @ 250% =	135.0 AMPS
3 H.P. MOTOR		9.6 AMPS
1 H.P. MOTOR		3.6 AMPS
⅙ H.P. MOTOR		0.7 AMPS
15 KVA HEATER	36.1 AMPS @ 125% =	45.0 AMPS
	TOTAL	194.0 AMPS

MAXIMUM FUSE — 200 AMPS

N.E.C. Section 430-58

Branch Circuit Breaker Size

EXAMPLE II
TABLE 430-152 LISTS THIS TYPE OF MOTOR FOR MAXIMUM INVERSE TIME CIRCUIT BREAKER SETTING OF 200% OF FULL LOAD CURRENT.

20 HP. MOTOR	54.0 AMPS @ 200% =	108.0 AMPS
3 H.P. MOTOR		9.6 AMPS
1 H.P. MOTOR		3.6 AMPS
⅙ H.P. MOTOR		0.7 AMPS
15 KVA HEATER	36.1 AMPS @ 125% =	45.0 AMPS
	TOTAL	167.9 AMPS

MAXIMUM CIRCUIT BREAKER — 175 AMPS

N.E.C. Section 430-62 (a)

Minimum Size Switch Or Circuit Breaker

20 HP. MOTOR	54.0 AMPS @ 115% =	62.1 AMPS
3 H.P. MOTOR		9.6 AMPS
1 H.P. MOTOR		3.6 AMPS
⅙ H.P. MOTOR		0.7 AMPS
15 KVA HEATER	36.1 AMPS @ 125% =	45.1 AMPS
	TOTAL	121.1 AMPS

MINIMUM FUSE — 125 AMPS
(40 H.P. RATED MINIMUM SWITCH)

20 HP. MOTOR	54.0 AMPS @ 115% =	62.1 AMPS
3 H.P. MOTOR		9.6 AMPS
1 H.P. MOTOR		3.6 AMPS
⅙ H.P. MOTOR		0.7 AMPS
15 KVA HEATER	36.1 AMPS @ 125% =	45.1 AMPS
	TOTAL	121.1 AMPS

MINIMUM CIRCUIT BREAKER — 125 AMPS

N.E.C. Section 430-72, Table 430-72 (b)

Overcurrent Protection Required For Control Circuit

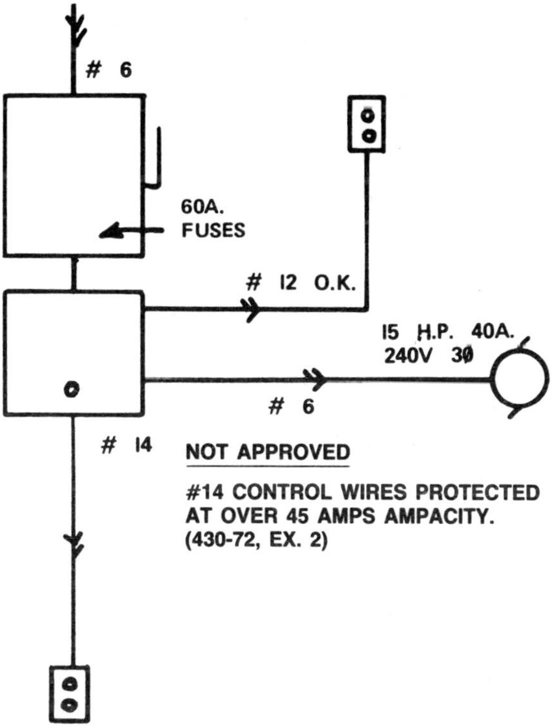

NOT APPROVED

#14 CONTROL WIRES PROTECTED AT OVER 45 AMPS AMPACITY. (430-72, EX. 2)

N.E.C. Section 430-72 (b), Ex. 2

Additional O.C.P. Not Required

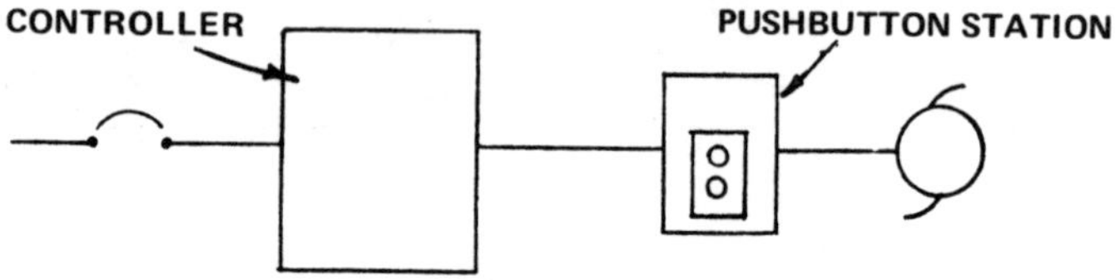

Branch circuit short circuit protective device rated not larger than the value specified in column "C" of Table 430-72(b) for control conductors which extend beyond the controller.

N.E.C. Section 430-72 (c), Ex. 3

Additional O.C.P. Not Required

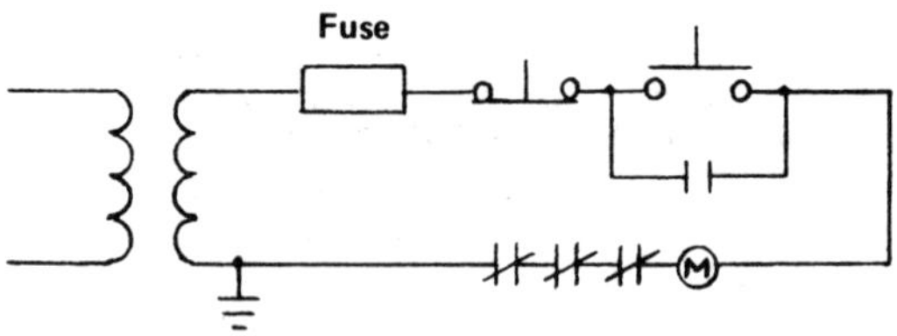

Where overcurrent protection is provided for the primary of a control transformer in accordance with Sec. 450-3.

N.E.C. Section 430-72 (b), Ex. 4

Additional O.C.P. Not Required

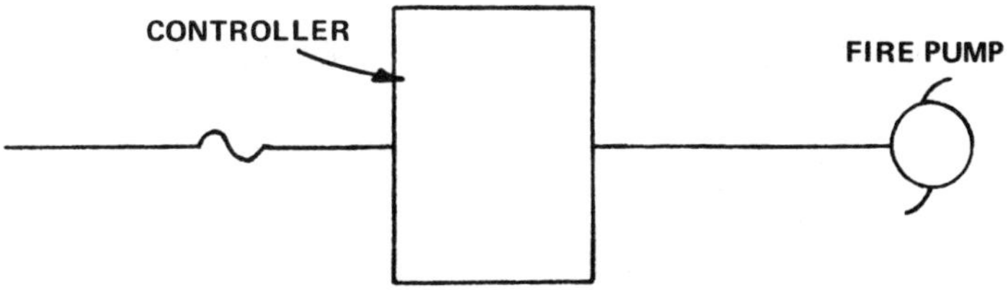

Where a hazard could exist
as in the case of a fire pump

N.E.C. Section 430-72 (b), Ex. 1

Overcurrent Protection For Motor Control Circuits

The branch-circuit protective device may serve as control-circuit conductor protection if . . .

. . . the control conductors do not extend beyond the controller enclosures. The START-STOP buttons are in the controller.

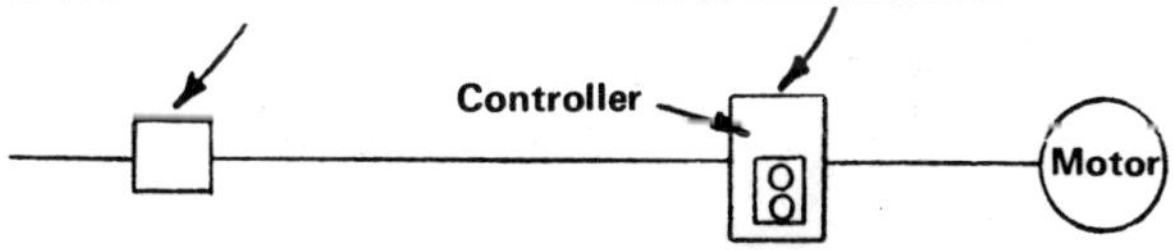

WHERE THE RATING OF THE PROTECTIVE DEVICE(S) IS NOT MORE THAN THE VALUE SPECIFIED IN COLUMN "B" OF TABLE 430-72(b)

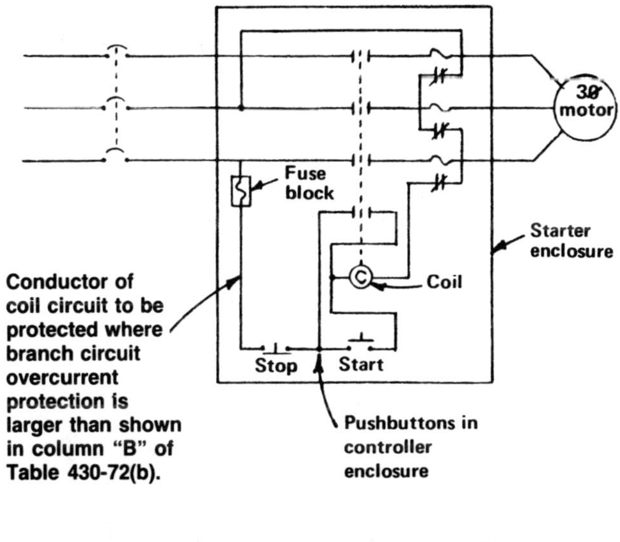

N.E.C. Sections 430-74 & 430-74 (a)

Motor Control Circuit Disconnect

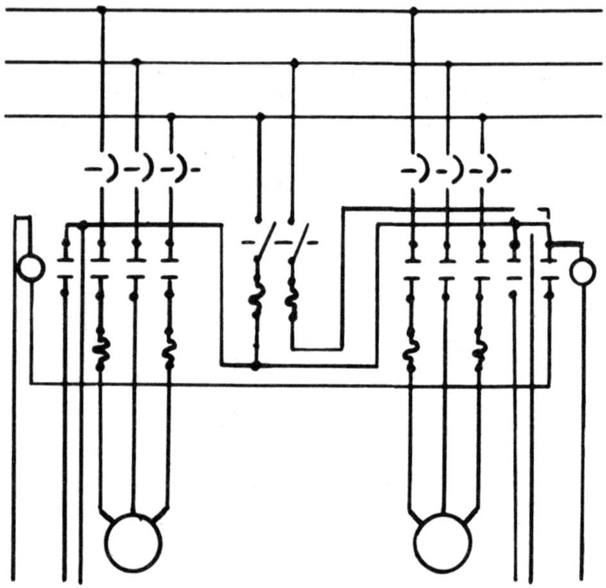

A SEPARATE SWITCH SHALL BE PROVIDED TO DISCONNECT CONTROL CIRCUIT WHERE MAIN DISCONNECT DOES NOT OPEN THE CONTROL CIRCUIT. SHALL BE LABELED WARNING, CONTROL CIRCUIT, ETC.

DISCONNECTING MEANS SHALL BE LOCATED ADJACENT TO EACH OTHER.

N.E.C. Section 430-102

Locking Means For Motor Controllers

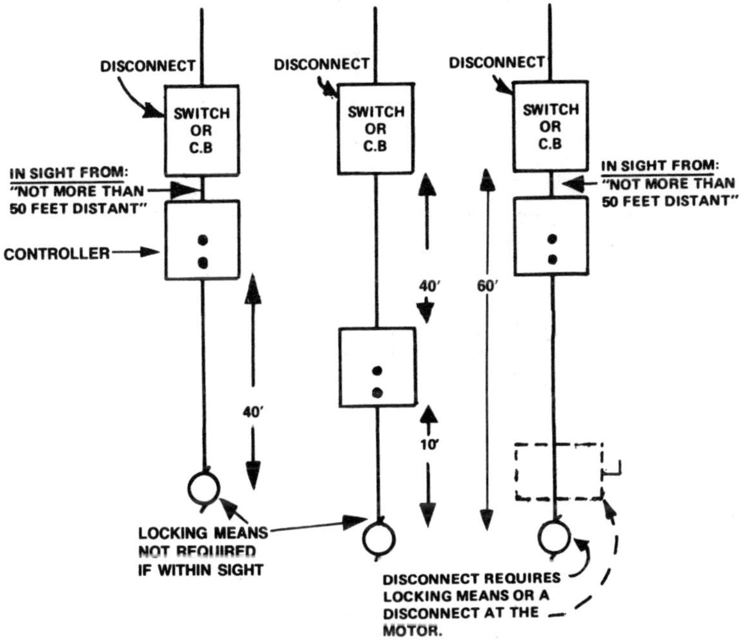

430.102 A DISCONNECTING MEANS SHALL BE LOCATED IN SIGHT FROM THE MOTOR CONTROLLER.

N.E.C. Section 430-102 (b)

Locking Means Required For Motor Disconnect

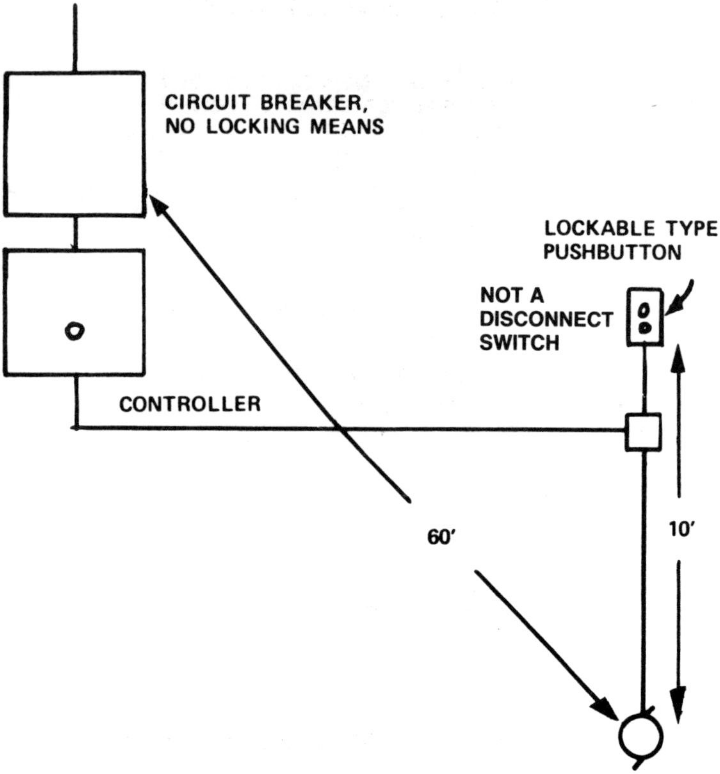

NOT APPROVED

DISCONNECT AND CONTROLLER NOT CAPABLE OF BEING LOCKED OPEN AND MORE THAN 50′ FROM MOTOR.
NOTE: THE REMOTE CONTROL STATION, LOCK–TYPE OR NOT IS NOT CONSIDERED THE CONTROLLER.

N.E.C. Section 430-109, Ex. 4

Disconnecting Means For Over 40 H.P. DC Or Motors Above 100 H.P. AC

DISCONNECTING MEANS MAY BE A GENERAL USE ISOLATION SWITCH MARKED "DO NOT OPEN UNDER LOAD"

OR

ACCEPTABLE:

SWITCH MARKED FOR THE HORSEPOWER.

OR

MANUFACTURER'S PUBLISHED DATA INDICATING THAT THE SWITCH IS CAPABLE OF INTERRUPTING A CURRENT OF NOT LESS THAN SIX (6) TIMES THE FULL LOAD CURRENT RATING OF THE MOTOR.

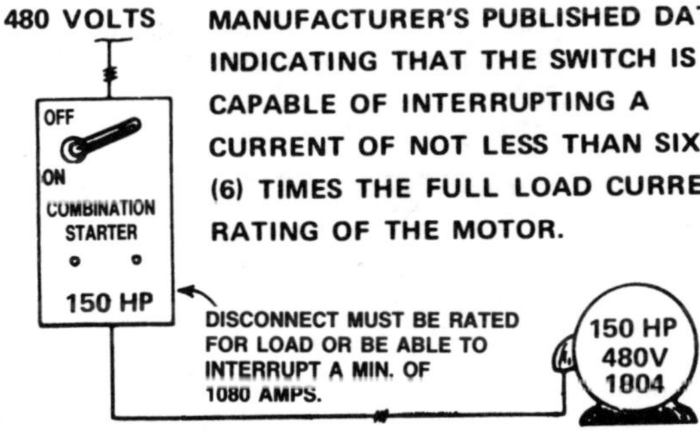

N.E.C. Section 430-109, Ex. 5

Cord And Plug Disconnect

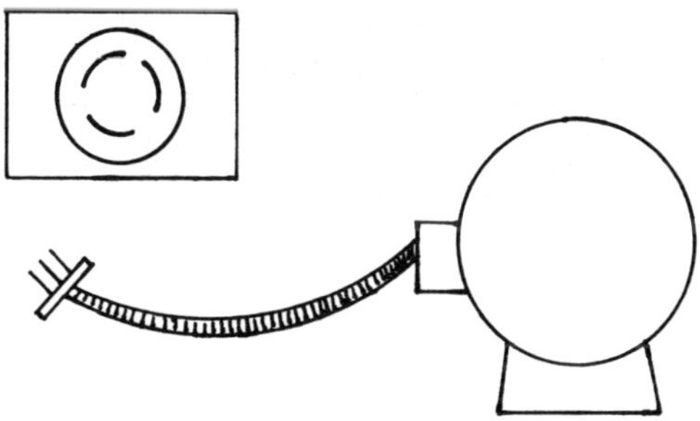

HORSEPOWER RATED PLUG REQUIRED IF USED AS DISCONNECT (NOT REQUIRED FOR APPLIANCES. SEE ART. 422-22 AND ROOM AIR CONDITIONERS. SEE SEC. 440-63)

N.E.C. Article 430, Part H

Disconnecting Means

N.E.C. Sections 430-110 & 430-110 (c)

Ampere Rating And Interrupting Capacity For Combination Loads

(1) The size of the disconnecting means shall be determined from the summation of all currents, including resistance loads, at the full-load condition and also at the locked-rotor condition. The current so obtained shall be considered as a single motor for this requirement as follows:

The full-load current of each motor selected from Table 430-150 shall be added to the amperes of the resistance load to obtain the full-load current for the combined load.

The locked-rotor current equivalent to the horsepower rating of each motor selected from Table 430-151 added to the amperes of the resistance load to obtain the locked-rotor current for the combined load.

EXAMPLE

The following 240V. 3ø equivalent will be served by a single disconnect switch:

2 - 25 H.P. Motors
2 - 5 H.P. Motors
1 - 21 K.W. Heater

Full-load current calculations:

2 - 25 H.P. Motors @ 68.0 Amps. each	136.0 Amperes
2 - 5 H.P. Motors @ 15.2 Amps. each	30.4 Amperes
1 - 21 K.W. Heater	50.6 Amperes
MINIMUM RATING	217.0 Amperes

It will be noted in Table 430-150 that the above current is between a 75 and a 100 H.P. motor rating. Therefore a 100 H.P. rated switch is required.

Locked-rotor current calculations:

2 - 25 H.P. Motor L.R.A. = 408 Amps. each	816.0 Amperes
2 - 5 H.P. Motor L.R.A. = 91 Amps. each	182.0 Amperes
1 - 21 K.W. Heater	50.6 Amperes
MINIMUM RATING	1048.6 Amperes

It will be noted in Table 430-151 that the above current is between a 60 and 75 H.P. motor rating. The higher of the two H.P. ratings, that which is the sum of the full-load current calculations, shall be used. This is equivalent to 100 H.P.

(2) The ampere rating of the switch shall not be less than 115 percent of the summation of all full-load currents in accordance with (c) (1) above.

Minimum ampere rating of the switch shall be 1.15 x 216.9 = 294.4 amperes. This requires a 400 ampere switch to be used.

N.E.C. Section 430-110

Disconnect Interrupting Capacity

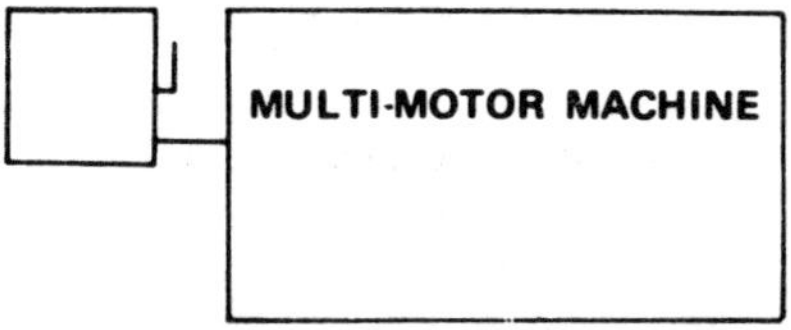

THE INTERRUPTING CAPACITY OF THE DISCONNECT SHALL NOT BE LESS THAN THE LOCKED-ROTOR CURRENT OF THE HIGHEST RATED MOTOR PLUS THE FULL LOAD CURRENT OF ALL OTHER OPERATING EQUIPMENT. THE DISCONNECT MAY NOT BE AN AMPERE RATED SWITCH OF THE NON-LOAD BREAK TYPE.

EXAMPLE:

* 1 - 50 HP 3Ø 240 V. CODE LETTER "R"
 1 - 20 HP 3Ø 240 V. CODE LETTER "R"
* 1 - 20 HP 3Ø 240 V. CODE LETTER "E"
* START TOGETHER

SECTION 430–7(b) LISTS LOCKED ROTOR AS FOLLOWS:

CODE LETTER "R" MAX. 15.99 KVA/HP
CODE LETTER "E" MAX. 4.99 KVA/HP

THUS:

50 HP X 15.99 KVA/HP X 2.41 AMPS/KVA	= 1927
20 HP X 4.99 KVA/HP X 2.41 AMPS/KVA	= 241
20 HP FULL LOAD AMPS	= 54
TOTAL REQUIRED INTERRUPTING CAPACITY OF DISCONNECT, AMPS	= 2222

(400 A. SWITCH X 600% = 2400 AMPS I.C.)

NOTE: LARGE MOTORS STARTING TOGETHER MAY CAUSE DISCONNECT PROBLEMS.

N.E.C. Section 430-110

Ampere Rating For Disconnecting Means

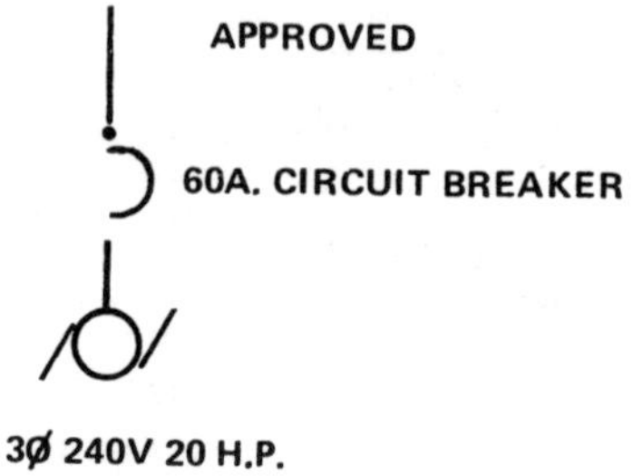

THE DISCONNECTING MEANS FOR A MOTOR SHALL HAVE AN AMPERE RATING OF NOT LESS THAN 115% OF THE FULL LOAD CURRENT RATING OF THE MOTOR, PROVIDED IT WILL START AND ACCELERATE THE LOAD.

N.E.C. Section 430-112 & 113

Disconnecting Means

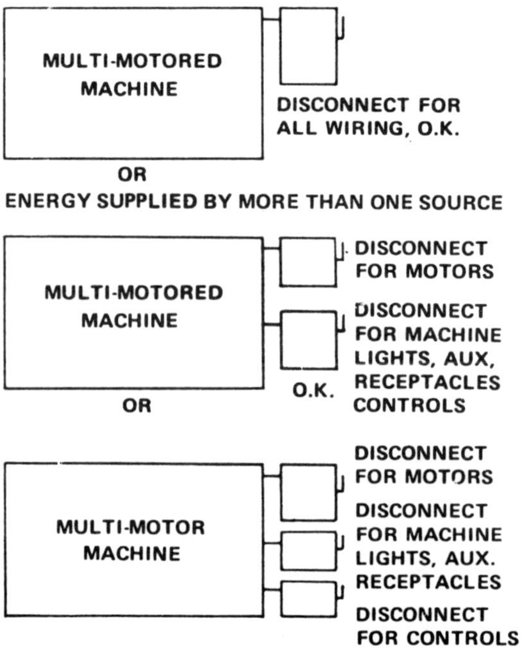

ALL DISCONNECTS SHALL BE PROPERLY MARKED TO INDICATE THE PART OF THE WIRING CONTROLLED BY EACH DISCONNECT.

N.E.C. Section 430-112

Single Disconnect For Multi-motored Machine

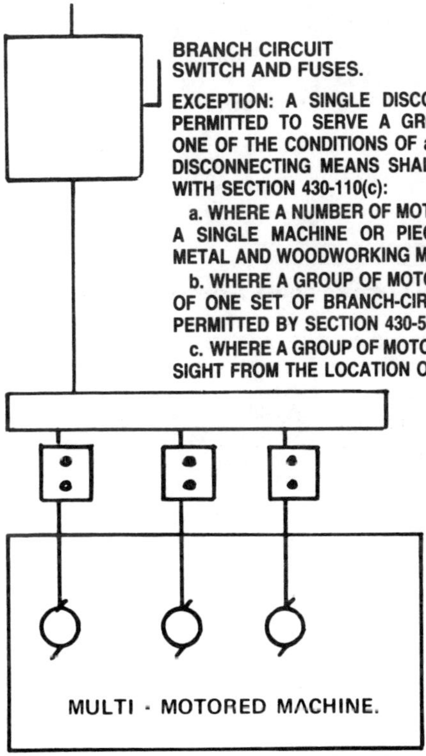

EXCEPTION: A SINGLE DISCONNECTING MEANS SHALL BE PERMITTED TO SERVE A GROUP OF MOTORS UNDER ANY ONE OF THE CONDITIONS OF a, b AND c BELOW. THE SINGLE DISCONNECTING MEANS SHALL BE RATED IN ACCORDANCE WITH SECTION 430-110(c):

a. WHERE A NUMBER OF MOTORS DRIVE SEVERAL PARTS OF A SINGLE MACHINE OR PIECE OF APPARATUS, SUCH AS METAL AND WOODWORKING MACHINES, CRANES AND HOISTS.

b. WHERE A GROUP OF MOTORS IS UNDER THE PROTECTION OF ONE SET OF BRANCH-CIRCUIT PROTECTIVE DEVICES AS PERMITTED BY SECTION 430-53(a).

c. WHERE A GROUP OF MOTORS IS IN A SINGLE ROOM WITHIN SIGHT FROM THE LOCATION OF THE DISCONNECTING MEANS.

N.E.C. Sections 430-141, 142, 143

Grounding Of Motors

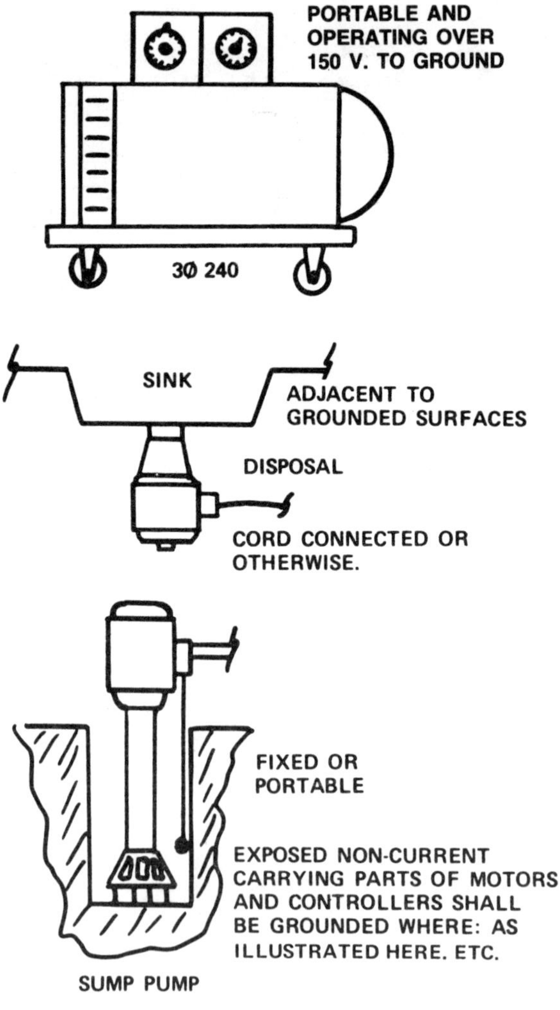

N.E.C. Section 430-144

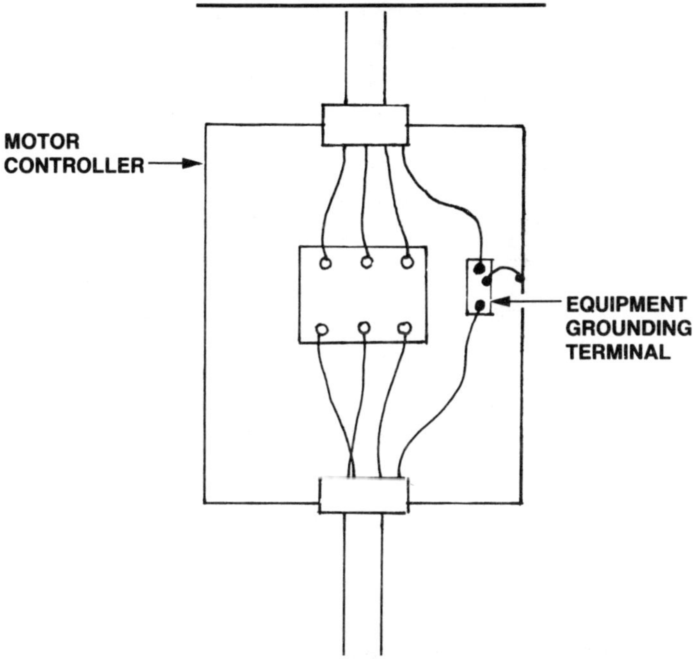

CONTROLLER ENCLOSURES SHALL HAVE MEANS FOR ATTACHMENT OF AN EQUIPMENT GROUNDING CONDUCTOR TERMINATION IN ACCORDANCE WITH SEC. 250-113.

N.E.C. Section 440-11

Location Of Emergency Stop Switch For Refrigeration Equipment

NOTE: THIS SECTION OF THE N.E.C. DOES NOT SPECIFICALLY ADDRESS THE PROBLEM OF EMERGENCY STOP SWITCHES FOR REFRIGERATION EQUIPMENT, WHICH IS COVERED BY SECTION 1509 OF THE UNIFORM MECHANICAL CODE, AS WELL AS BY OTHER LOCAL CODES. THE FOLLOWING DIAGRAMS ILLUSTRATE GOOD PRACTICE TO COMPLY WITH THESE CODES.

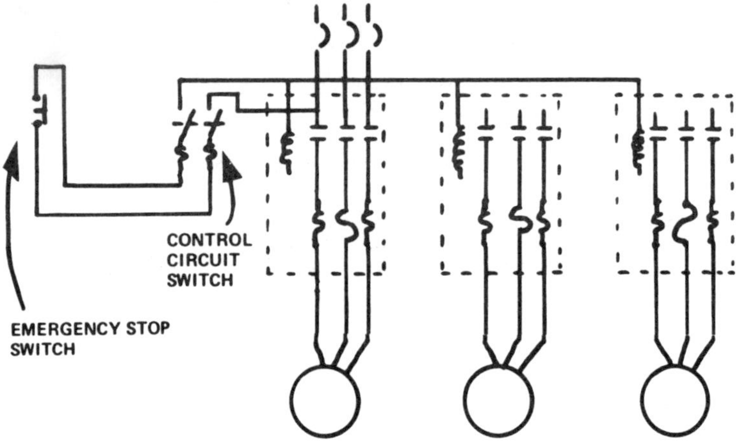

REFRIGERATION EQUIPMENT IN MACHINE ROOM

EMERGENCY STOP SWITCH IS REQUIRED TO BE INSTALLED OUTSIDE OF, AND WITHIN 10′, OF DOOR TO STOP ALL ELECTRICALLY OPERATED MACHINERY LOCATED IN A *REFRIGERATION MACHINERY ROOM,* EXCEPT THE EXHAUST VENTILATION SYSTEM.

Refrigeration Machinery Rooms

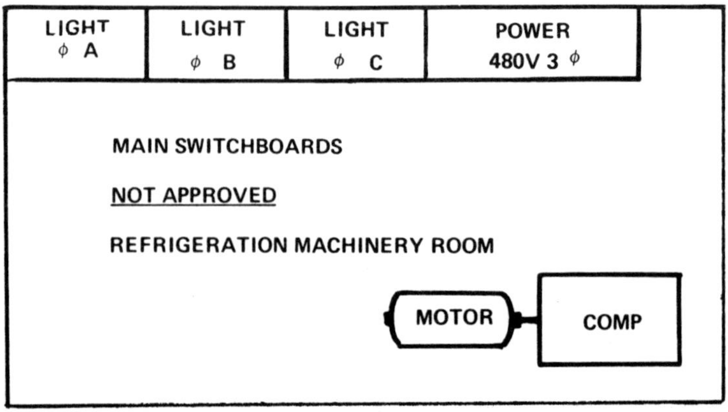

A SEPARATE MACHINE ROOM FOR A SYSTEM CONTAINING: 100 HP OR MORE WITH CLASS 1 REFRIGERANT OR CLASS 2 REFRIGERANT. NO MINIMUM H.P.

N.E.C. Section 440-14

Disconnecting Means For Air Conditioning Units

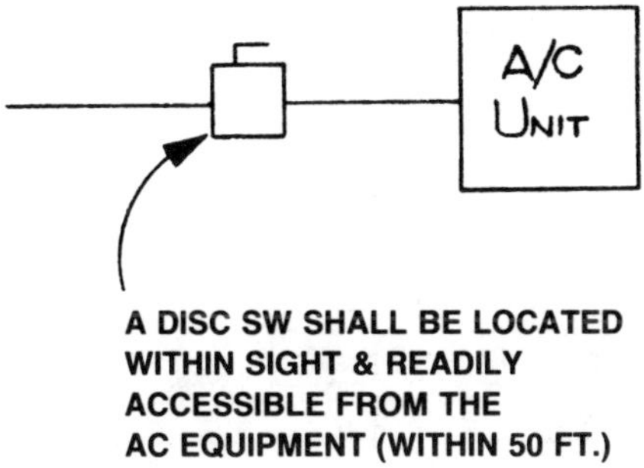

N.E.C. Section 440-41

Computing Controller Size For Hermetically Sealed Refrigeration Units

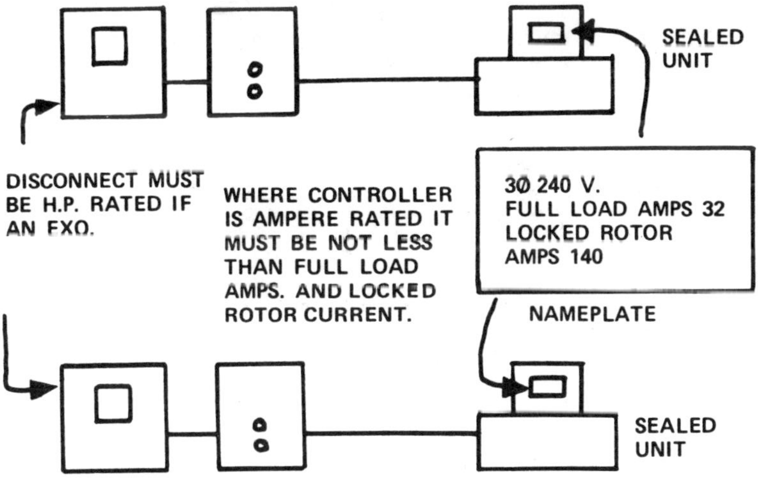

WHERE CONTROLLER IS
H.P. RATED, SEE BELOW:

TABLE 430-150 FULL LOAD AMPS	TABLE 430-151
10 H.P. = 27 AMPS	132 AMPS I.C. = 7½ H.P.
? H.P. = 32 AMPS	140 AMPS I.C. = ? H.P.
15 H.P. = 40 AMPS	162 AMPS I.C. = 10 H.P.

USE THE RATING FROM TABLE 430-150 OR TABLE 430-151 THAT INDICATES THE LARGER RATING. (IN THIS CASE TABLE 430-150 INDICATES A SIZE OF ABOUT 11.9 H.P. THE CONTROLLER MUST BE AT LEAST 11.9 H.P. THE STANDARD SIZE WOULD BE 15 H.P.)

N.E.C. Section 445-4 (a)

Overcurrent Protection For Generators

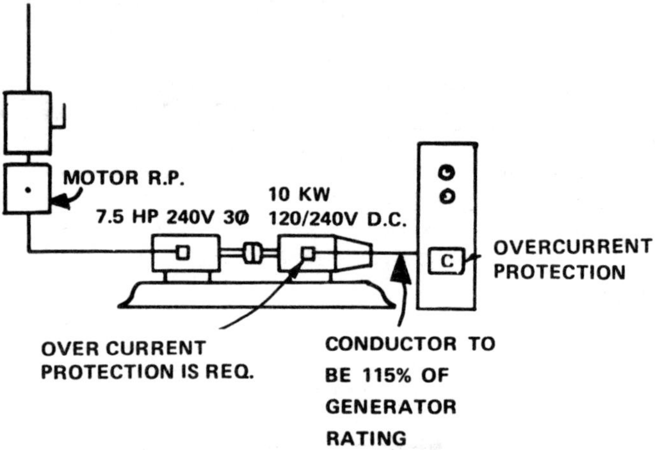

C.B. PROTECTING GENERATOR IS REQUIRED EVEN THOUGH H.P. OF MOTOR IS OF SIZE TO PROTECT GENERATOR

N.E.C. Section 445-4 (c)

Overcurrent Protection For Generators

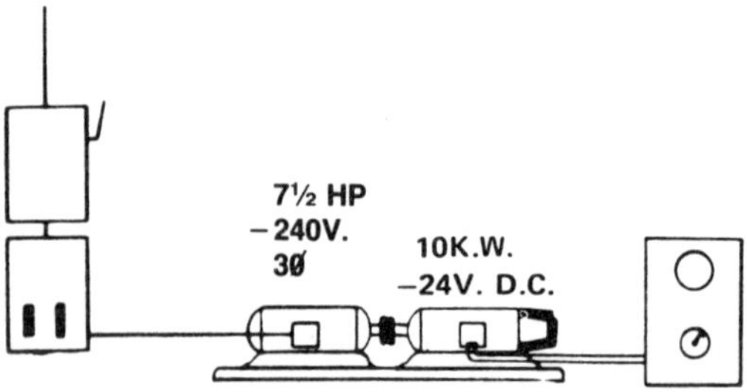

O.K. WITHOUT SEPARATE OVERCURRENT PROTECTION FOR GENERATOR - UNDER 65 V. &

PROTECTED BY RUNNING PROTECTION OF MOTOR, WHERE THE DEVICE OPERATES WHEN GENERATOR IS DELIVERING NOT MORE THAN 150% OF ITS FULL LOAD RATED CURRENT.

N.E.C. Section 450-3 (a)

Protection Of High Voltage (Over 600V) Transformer Depends On Conditions At The Place Of Installation

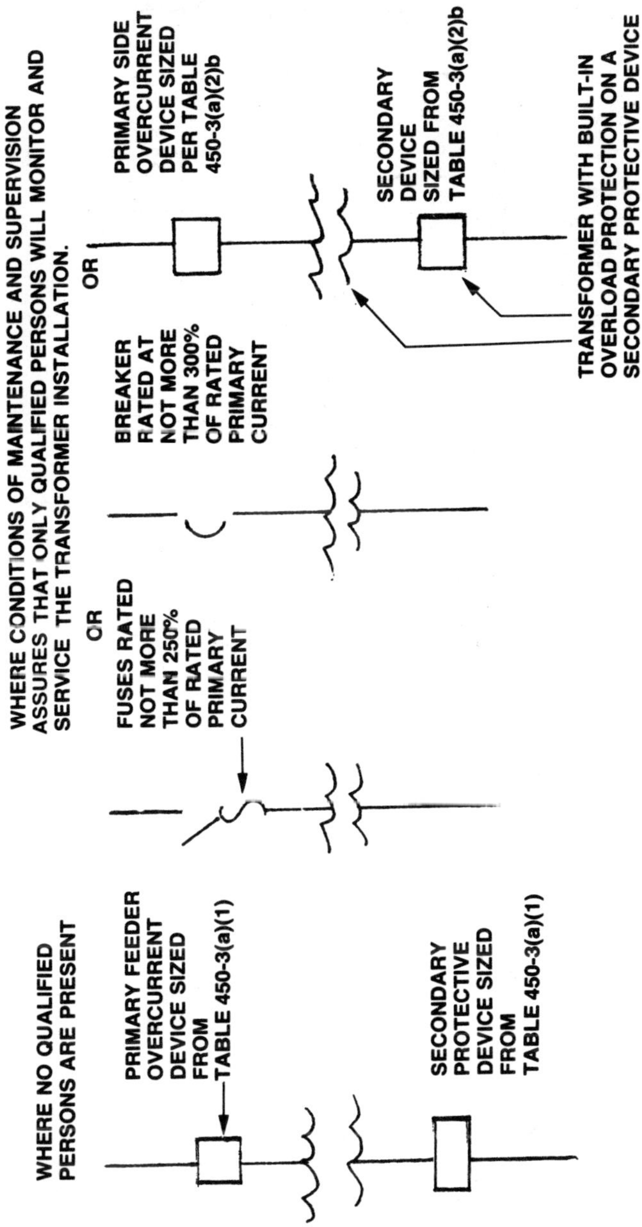

N.E.C. Section 450-3 (b)

Overcurrent Protection For Transformer Primary

PRIMARY OF TRANSFORMERS SHALL BE PROTECTED AT NOT TO EXCEED 125% OF THE RATED PRIMARY CURRENT OF THE TRANSFORMER.

EXCEPTIONS:

1. PRIMARY IS RATED 9 AMPS OR MORE AND 125% O.C.P. DEVICE IS NOT STANDARD. THE NEXT SIZE HIGHER RATING IS PERMITTED. SEE SEC. 240-6 FOR STANDARD RATINGS.

2. WHERE PRIMARY IS RATED LESS THAN 9 AMPS, PRIMARY SHALL BE PROTECTED AT NOT TO EXCEED 167% OF THE PRIMARY RATING. PAGE 44 (a)

WHERE PRIMARY IS RATED LESS THAN 2 AMPERES, THE PRIMARY SHALL BE PROTECTED AT NOT TO EXCEED 300% OF THE PRIMARY RATING.

3. WHERE SECONDARY SIDE OF THE TRANSFORMER RATED AT NOT MORE THAN 125% OF THE SECONDARY RATING, THE PRIMARY OVERCURRENT PROTECTIVE DEVICE SHALL NOT EXCEED 250% OF THE PRIMARY RATING.

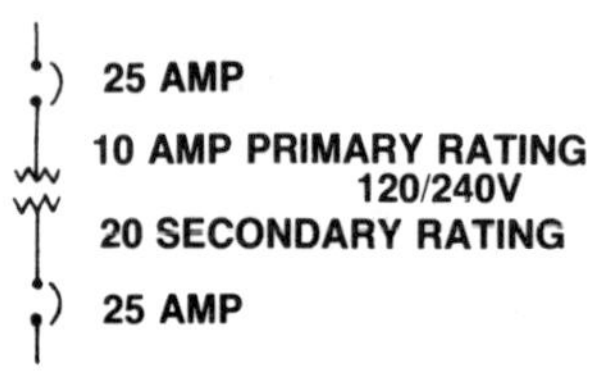

N.E.C. Section 450-45 (b) & (c)

Transformer Ventilation

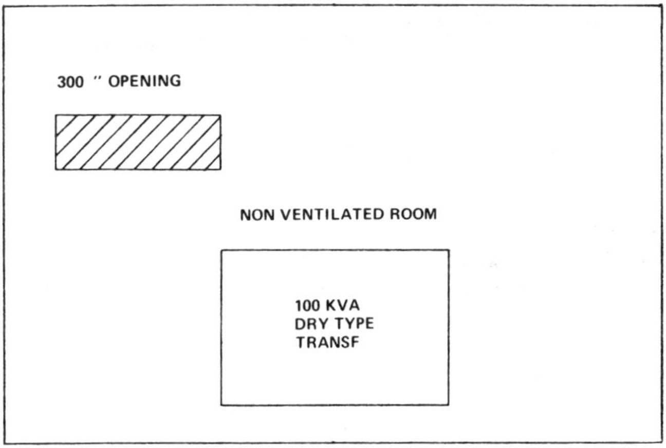

VAULT VENTILATION MAY BE PROVIDED BY 1 OR MORE OPENINGS IN OR NEAR THE ROOF.

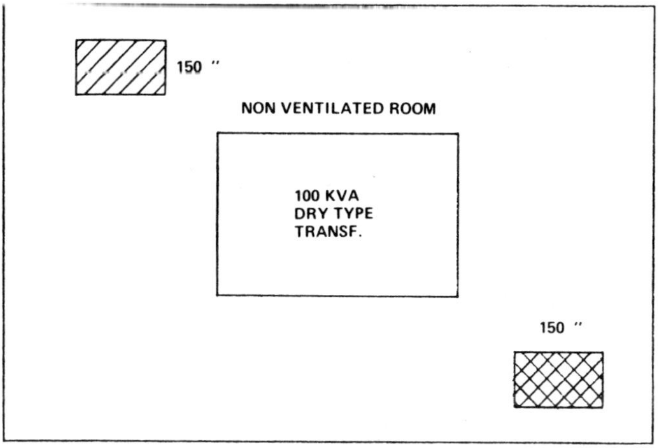

(b) VENTILATING OPENINGS MAY HAVE ROUGHLY 1/2 OF THE TOTAL AREA OF OPENINGS NEAR THE FLOOR & REMAINDER IN THE ROOF OR SIDEWALLS NEAR THE ROOF.

(c) VENTILATING OPENINGS SHALL BE NOT LESS THAN 3 SQ. IN. PER KVA NET AREA EXCLUDING ANY AREA OCCUPIED BY SCREENS, GRATING OR LOUVERS WITH A *MINIMUM of 1 SQUARE FOOT* FOR ANY CAPACITY UNDER 50 KVA.

N.E.C. Sections 460-8 & 460-9

Capacitors On Motor Circuits

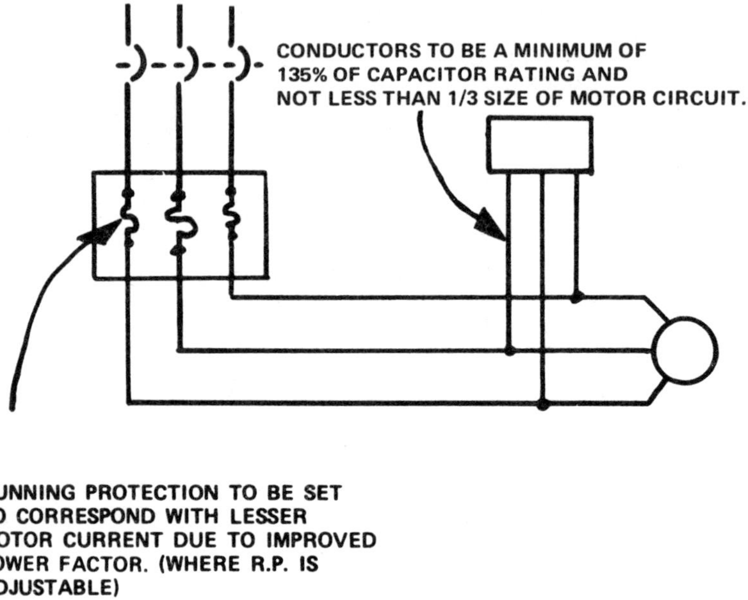

N.E.C. Section 460-8 (a)

Conductor Size And Overcurrent Protection For Capacitor

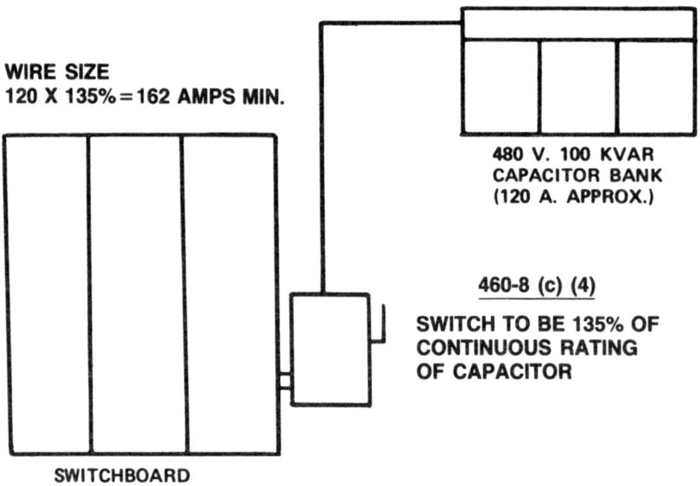

N.E.C. Section 500-2

Pressurized Rooms

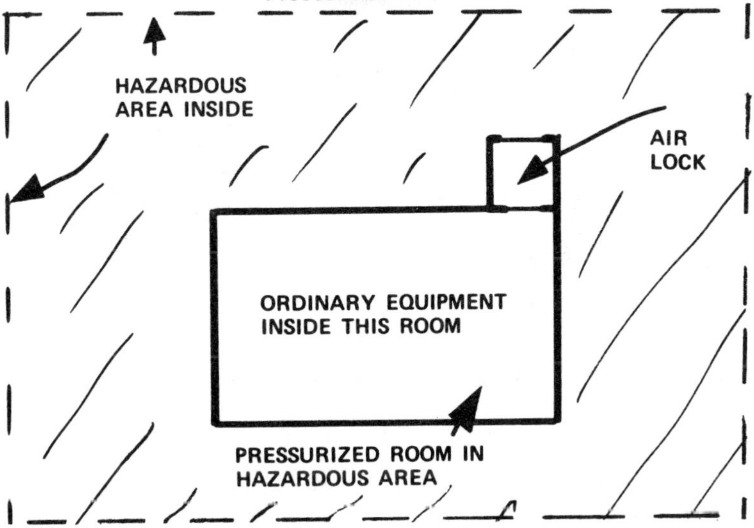

ROOM MAY BE CLASSIFIED AS A NON-HAZARDOUS AREA WHERE CONSTRUCTED AND EQUIPPED AS PRESCRIBED IN CODE.

N.E.C. Section 500-2

Approval Required For Specific Use

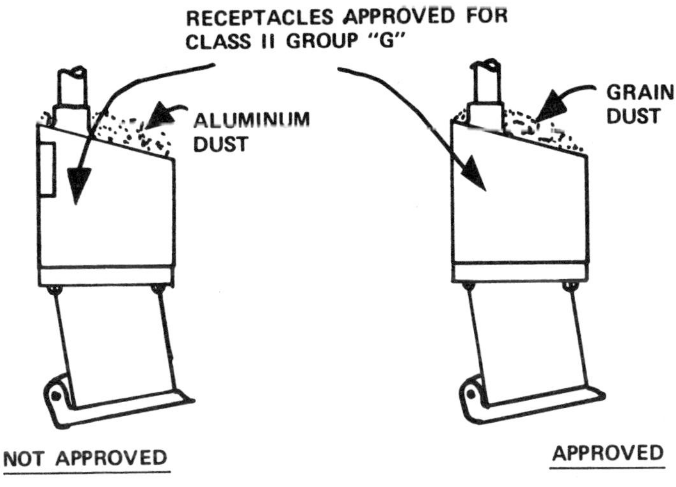

APPROVAL FOR GROUP "G" DUSTS INCLUDES GRAIN DUSTS, BUT DOES NOT INCLUDE ALUMINUM DUST

N.E.C. Section 501-4

Wiring Methods Class 1, Division 1 — Hazardous Location

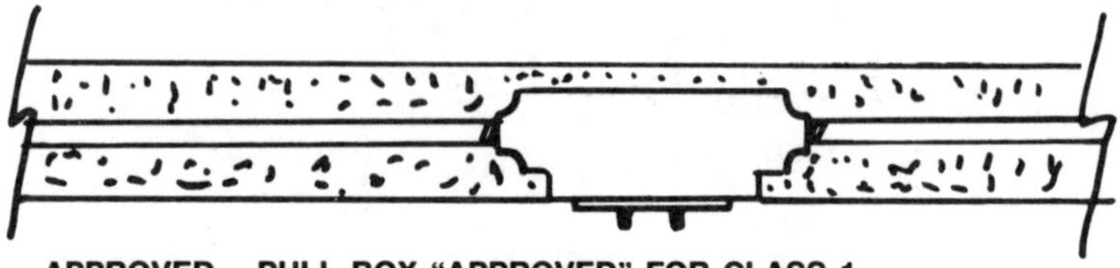

APPROVED – PULL BOX "APPROVED" FOR CLASS 1 DIVISION 1 HAZARDOUS LOCATION (AND GROUP VARIOUS CLASSIFICATION) WITH COVER OPENING IN AREA.

NOT APPROVED –K.O. TYPE PULL BOX WITH COVER OPENING INTO A CLASS 1 HAZARDOUS AREA.

N.E.C. Section 501-4 (a)

Wiring Method Class 1 Division 1 Hazardous Locations

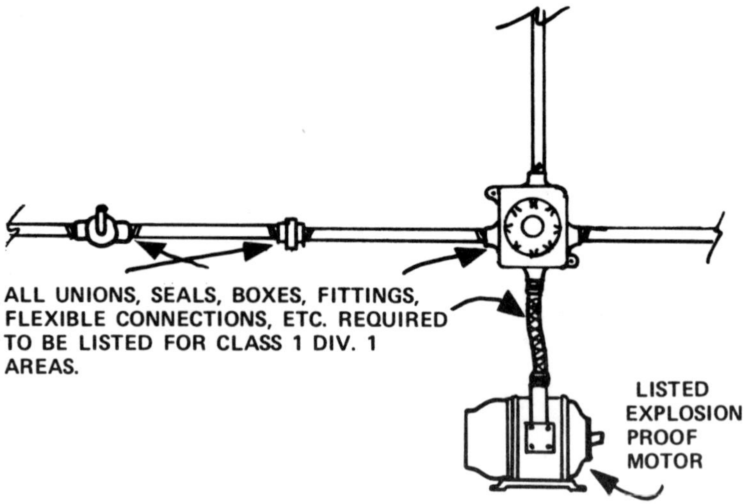

N.E.C. Section 501-4 (b)

Wiring Method Class 1 Division 2

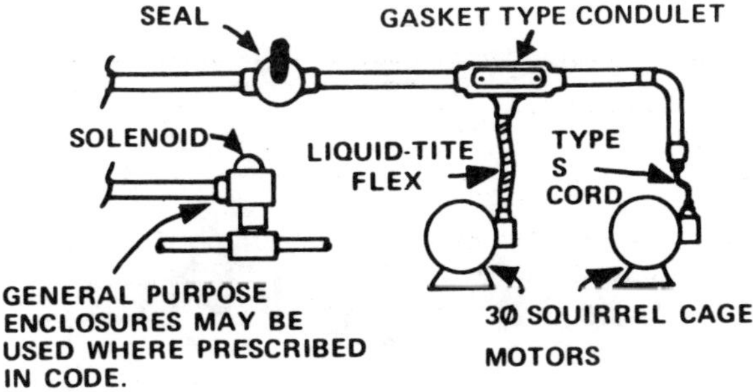

N.E.C. Section 501-5 (a)

Seals Class 1 Division 1

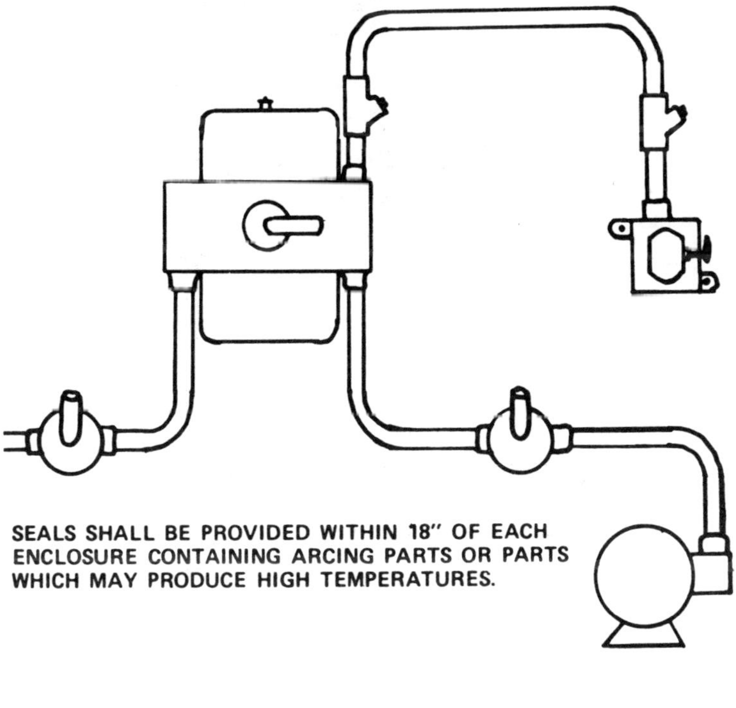

N.E.C. Section 501-5 (a) 4

Location Of Seals In Class 1 Locations

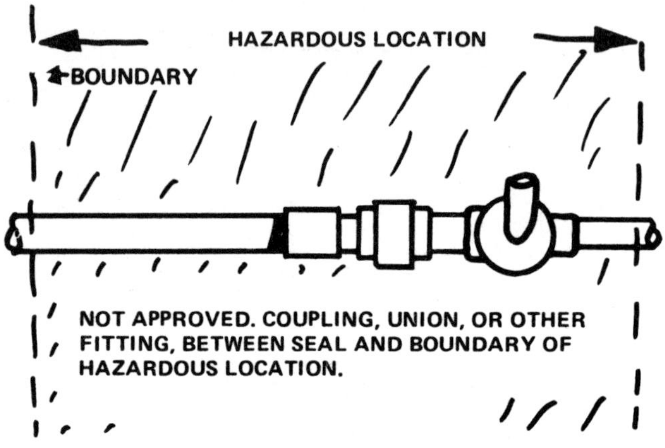

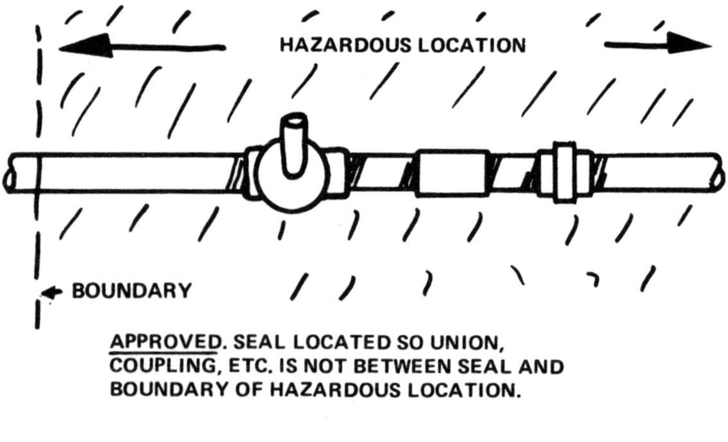

N.E.C. Section 501-5 (a) 4, (b) 2

Seals Required At Area Boundaries

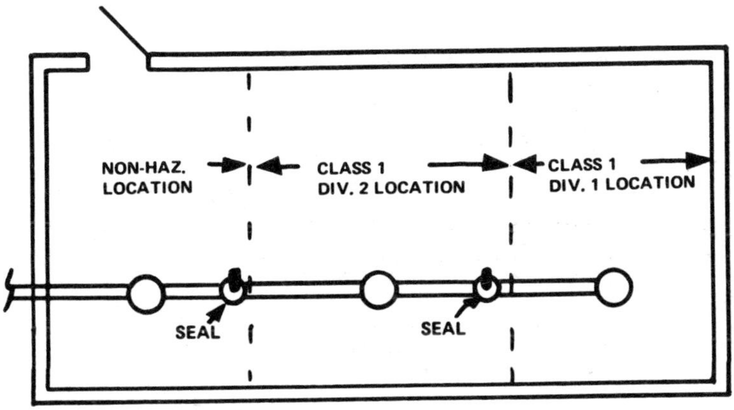

SEAL REQUIRED: WHERE CONDUIT LEAVES CLASS 1 DIV 1 LOCATION. AND WHERE CONDUIT LEAVES CLASS 1 DIV 2 LOCATION

N.E.C. Section 501-5 (b) 1

Seals Class 1 Division 2

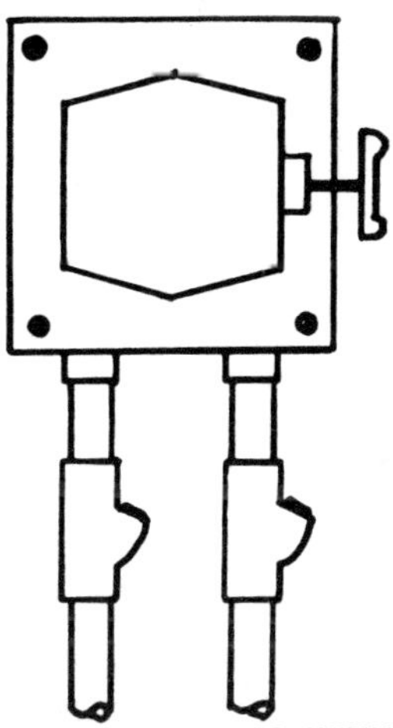

SEALS SHALL BE PROVIDED WITHIN 18" OF ENCLOSURES THAT ARE REQUIRED TO BE EXPLOSION-PROOF.

N.E.C. Section 501-5 (c)

Means For Sealing Class 1 Locations

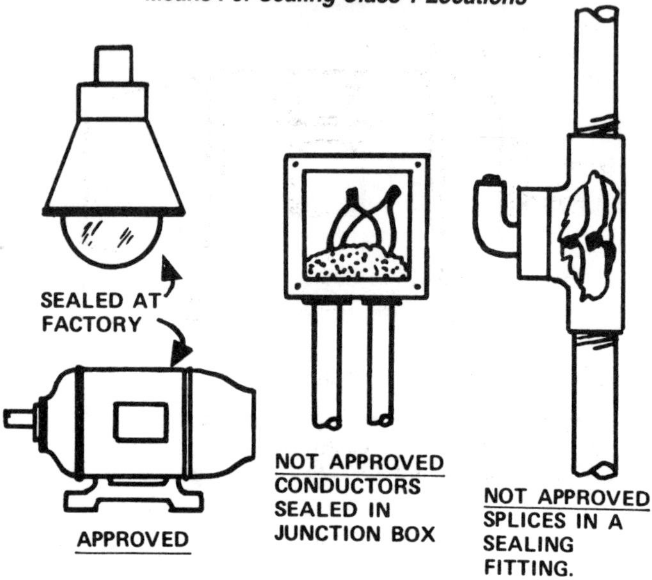

N.E.C. Section 501-6 (a) (b)

Types Of Switches And Circuit Breakers For Class 1 Areas

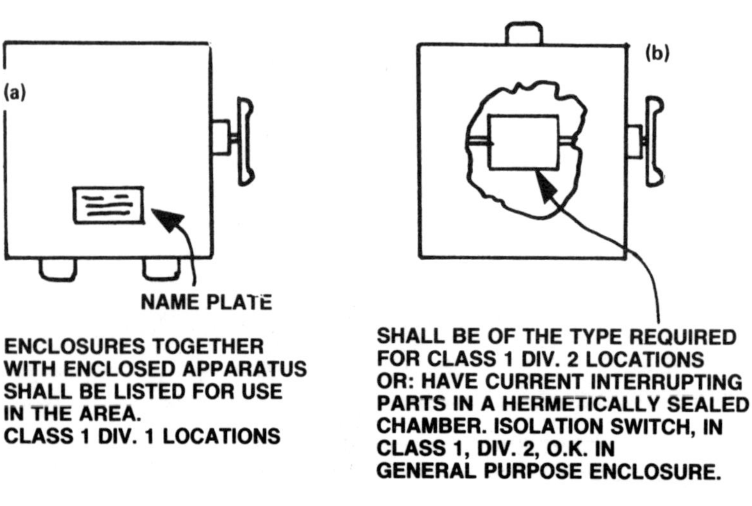

N.E.C. Section 501-8 (a)

Motors And Generators Class 1 Division 1

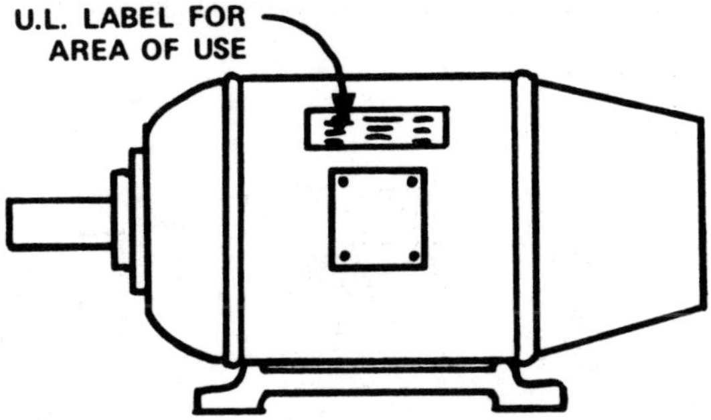

SHALL BE LISTED FOR THE LOCATION.

EXCEPT: TOTALLY ENCLOSED MOTORS DESIGNED FOR THE PURPOSE AND PROPERLY VENTILATED MAY BE USED.

N.E.C. Section 501-9 (a)

Fixtures And Supports Class 1 Division 1

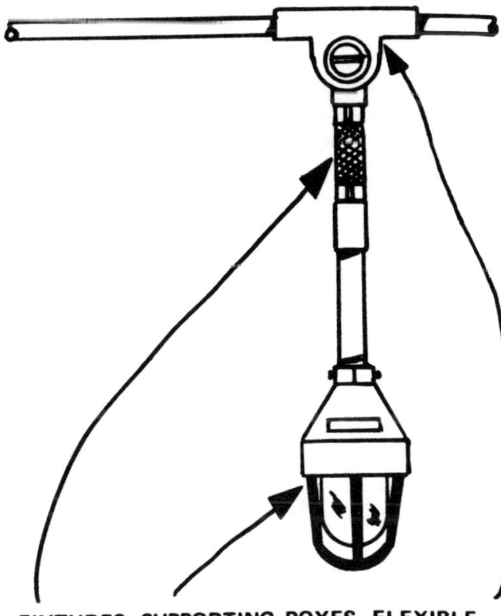

FIXTURES, SUPPORTING BOXES, FLEXIBLE STEMS, ETC. ALL LISTED IN U.L. "REDBOOK" UNDER "FIXTURES AND FIXTURE FITTINGS."

N.E.C. Sections 501-9 (a) 3 & 502-11 (a) 3

Pendant Fixture In Hazardous Area

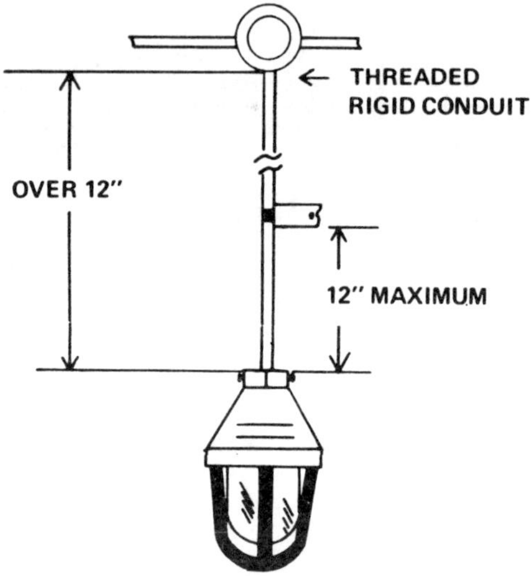

FOR STEMS OVER 12", PERMANENT & EFFECTIVE BRACING AGAINST LATERAL DISPLACEMENT SHALL BE PROVIDED AT A LEVEL NOT MORE THAN 12" ABOVE LOWER END OF STEM OR APPROVED FLEXIBLE FITTING OR CONNECTOR MAY BE INSTALLED WITHIN 12" OF SUPPORTING BOX.

N.E.C. Section 501-9 (b)

Lighting Fixtures Class 1 Division 2 Hazardous Areas

PORTABLE LAMPS

SHALL BE LISTED FOR AREA

OR

BE OF A SUITABLE ENCLOSED GASKETED TYPE

LIGHTING FIXTURES

IF SURFACE TEMP. REACHES 80% OF IGNITION TEMP. FIXTURE MUST BE LISTED FOR CLASS 1, DIV. 1

FIXED LIGHTING FIXTURES SHALL BE LISTED FOR AREA, OR BE OF A SUITABLE ENCLOSED AND GASKETED TYPE.

N.E.C. Sections 501-14, 502-14 & 503-12

Low Energy Systems Are Not Exempt

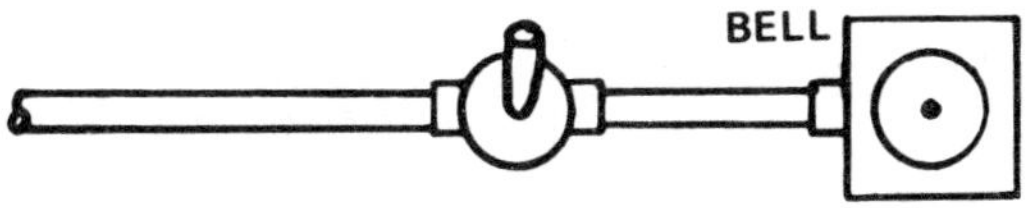

SIGNAL, ALARM, REMOTE CONTROL, AND COMMUNICATION SYSTEMS ARE NOT EXEMPT FROM CODE DUE TO LOW VOLTAGE OR WATTAGE, EXAMPLE ABOVE – 12 V., 1.6 A, BELL IN CLASS I DIV 1 AREA SHALL BE "APPROVED" FOR THE AREA AND WIRED IN RIGID CONDUIT OR MI CABLE, WITH SEALS, ETC.

N.E.C. Sections 501-16 (a) & 502-16 (a)

Bonding In Hazardous Areas

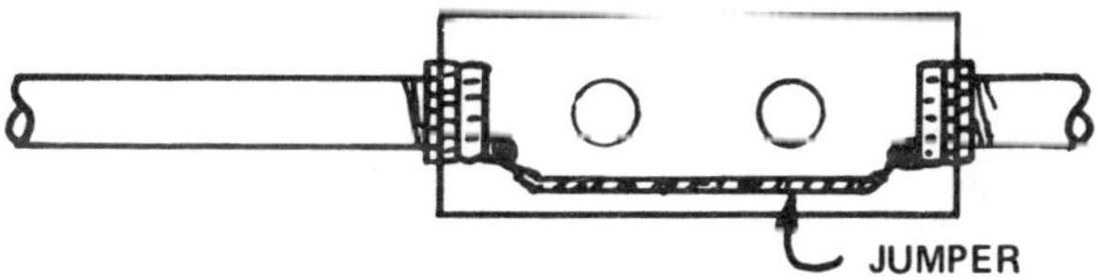

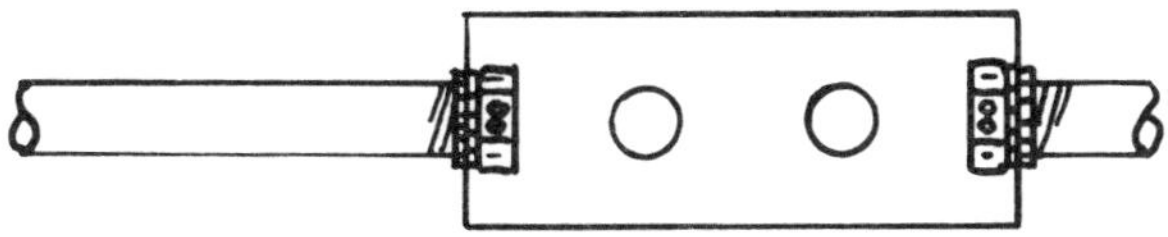

BONDING JUMPERS OR OTHER APPROVED MEANS SHALL BE USED TO INSURE GROUNDING CONTINUITY AROUND LOCKNUTS AND BUSHINGS.

N.E.C. Section 501-16 (a)

EVERY CONDUIT TERMINATION BACK TO SERVICE GROUNDING ELECTRODE MUST BE OF A SERVICE-TYPE BONDING CONNECTION

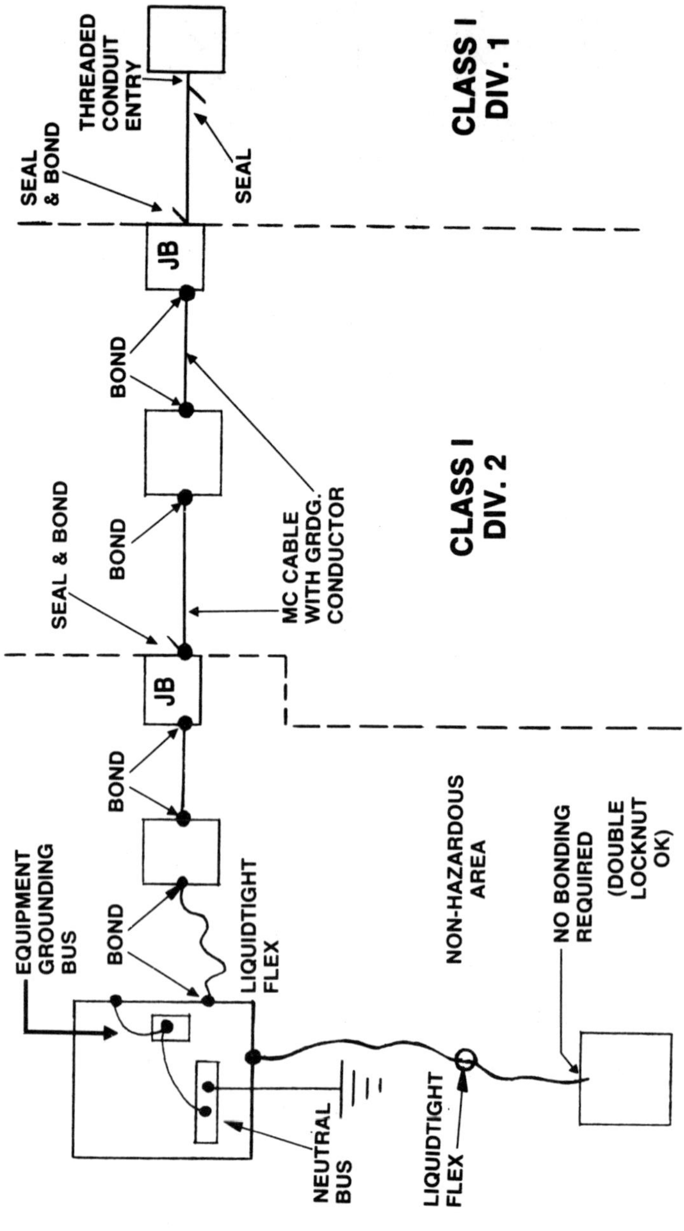

N.E.C. Section 502-4 (a)

Wiring Methods Class II Division 1

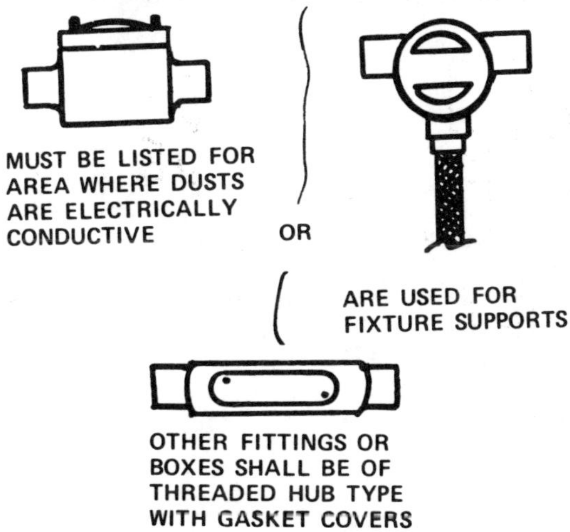

N.E.C. Section 502-4 (a)

Wiring Method Class II Division 1

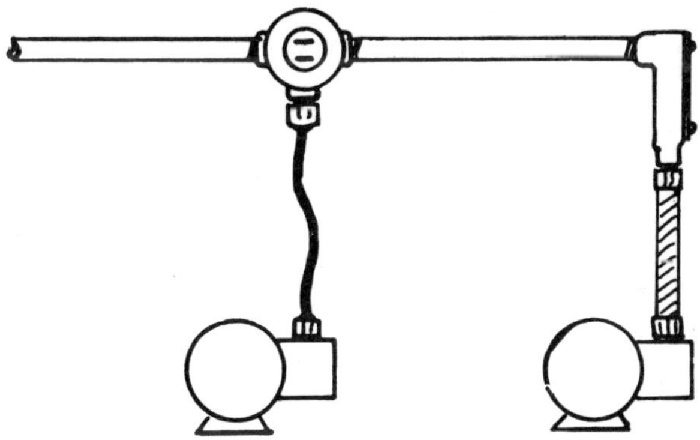

WIRING SHALL BE IN RIGID CONDUIT, INTERMEDIATE METAL CONDUIT OR MI CABLE:

EXCEPT: TYPE "S" CORD MAY BE USED WHERE FLEXIBILITY IS REQUIRED.

EXCEPT: WHERE FLEXIBILITY IS REQUIRED AND DUSTS ARE NOT ELECTRICALLY CONDUCTING, LIQUIDTIGHT FLEXIBLE CONDUIT MAY BE USED.

N.E.C. Section 502-4 (b)

Wiring Method Class II Division 2

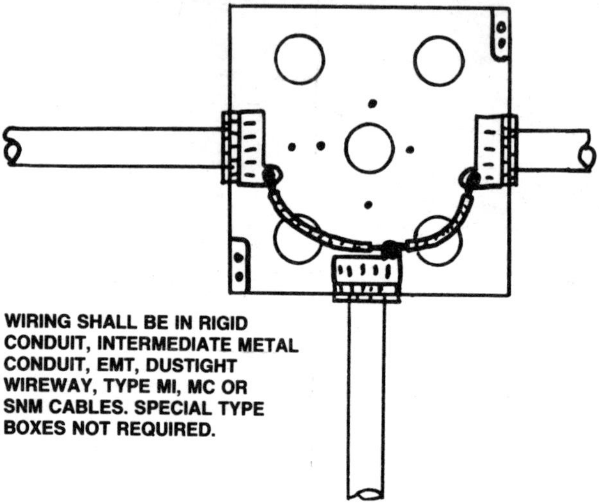

WIRING SHALL BE IN RIGID CONDUIT, INTERMEDIATE METAL CONDUIT, EMT, DUSTIGHT WIREWAY, TYPE MI, MC OR SNM CABLES. SPECIAL TYPE BOXES NOT REQUIRED.

N.E.C. Section 502-5

Sealing In Class II, Division 1 & 2

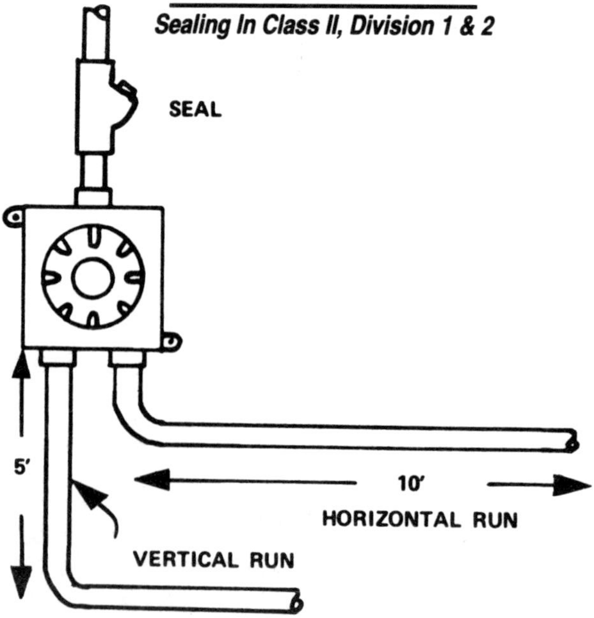

MEANS SHALL BE PROVIDED TO PREVENT THE COMMUNICATION OF DUST BETWEEN ENCLOSURES THAT ARE REQUIRED TO BE DUST-IGNITION-PROOF AND THOSE NOT REQUIRED TO BE DUST-TIGHT.

N.E.C. Section 502-6 (a)

Switches, Circuit Breakers, Etc., In Class II Division 1

SWITCHES, CIRCUIT BREAKERS, CONTROLLERS, AND FUSES SHALL BE PROVIDED WITH DUST-IGNITION-PROOF ENCLOSURES LISTED FOR CLASS II LOCATIONS.

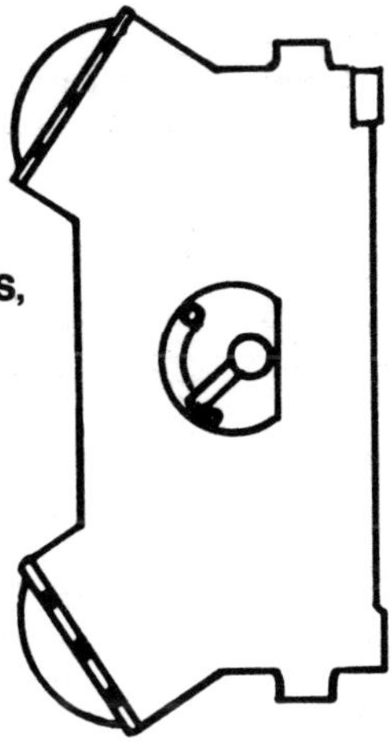

N.E.C. Sections 502-6 (b)

Switches, Circuit Breakers, Etc., For Class II Division 2

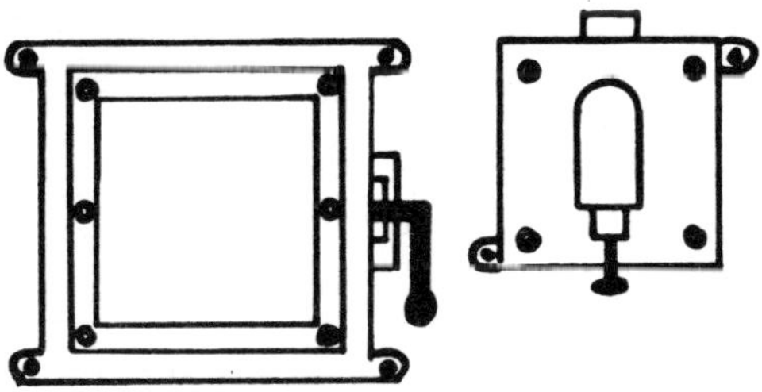

ENCLOSURES SHALL BE LISTED FOR CLASS II LOCATIONS; OR SHALL BE OF SUITABLE CAST ENCLOSURES WITH THREADED HUBS OR BOSSES, GASKETED COVERS AND NO HOLES FOR MOUNTING, AND SHALL BE DUSTIGHT.

N.E.C. Section 502-8 (a)

Motors Class II Division 1

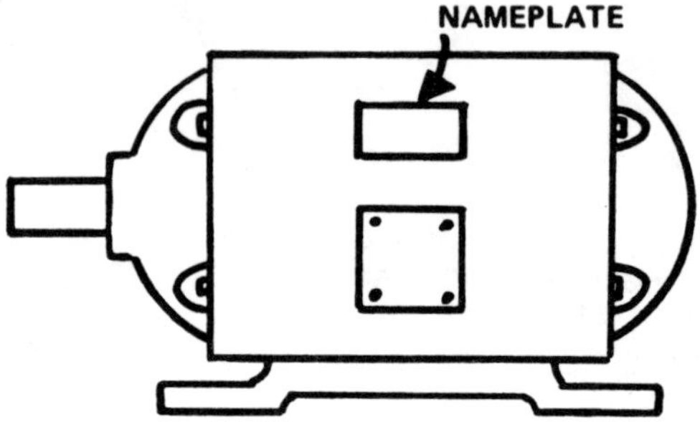

MOTORS FOR CLASS II DIV. 1 AREAS SHALL BE APPROVED FOR THE LOCATION OR BE TOTALLY ENCLOSED PIPE VENTILATED.

N.E.C. Section 502-8 (b)

Motors Class II Division 2

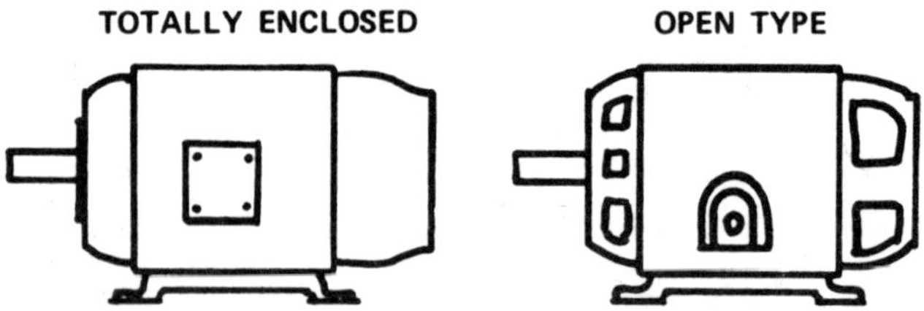

MOTORS FOR CLASS II DIV. 2 HAZARDOUS AREAS SHALL BE APPROVED FOR CLASS II LOCATIONS OR TOTALLY ENCLOSED NONVENTILATED, TOTALLY ENCLOSED PIPE VENTILATED OR TOTALLY ENCLOSED FAN COOLED.

EXCEPT: WHERE READILY ACCESSIBLE FOR ROUTINE CLEANING, TOTALLY ENCLOSED TYPES OR OPEN TYPES WITHOUT CONTACTS OR INTEGRAL RESISTANCES MAY BE USED.

N.E.C. Section 502-11 (a)

Lighting Fixtures Class II Division 1

LIGHTING FIXTURES AND SUPPORTING MEANS SHALL BE LISTED FOR CLASS II LOCATIONS. FIXTURES SHALL BE MARKED TO INDICATE MAXIMUM LAMP WATTAGE.

PORTABLE LAMPS SHALL BE LISTED FOR CLASS II LOCATIONS.

N.E.C. Section 502-11 (b)

Lighting Fixtures Class II Division 2

LIGHTING FIXTURES SHALL PROVIDE ENCLOSURES FOR LAMPS AND LAMP HOLDERS.

PORTABLE LAMPS SHALL BE APPROVED FOR CLASS II LOCATIONS.

N.E.C. Section 503-3 (a)

Class III

APPROVED WIRING METHODS: RIGID METAL CONDUIT, RIGID NONMETALLIC CONDUIT, INTERMEDIATE METAL CONDUIT, EMT, DUSTIGHT WIREWAYS OR TYPE MI, MC OR SNM CABLES.

N.E.C. Section 511-2

Hazardous Location In Garages

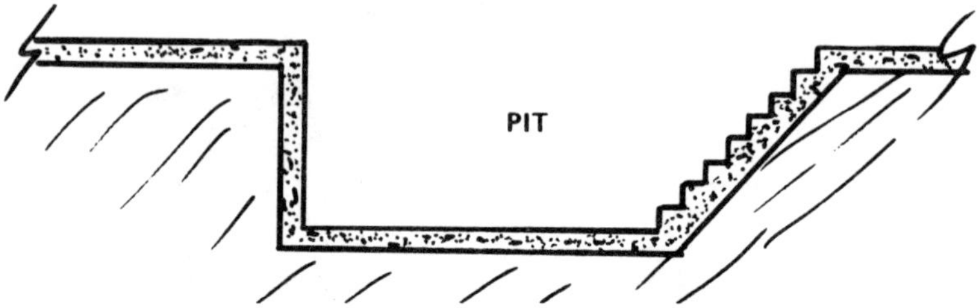

POORLY VENTILATED PIT, CLASS I DIV 1.

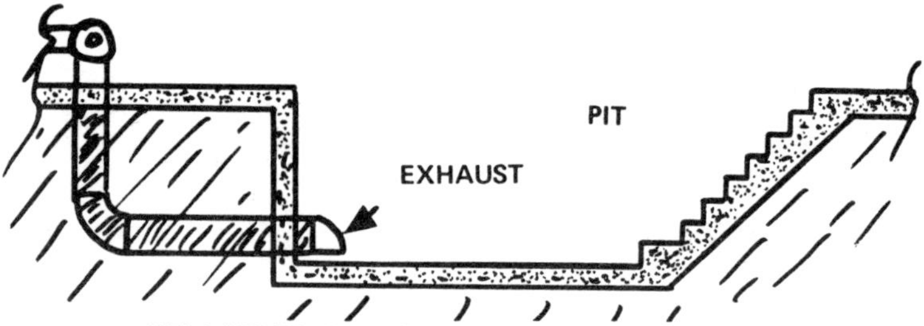

WELL VENTILATED PIT, CLASS I, DIV 2,
(NOT LESS THAN 6 AIR CHANGES PER HOUR)

NOTE: BASEMENT GARAGE AREAS AND OTHER POORLY VENTILATED GARAGE AREAS ARE CLASS I DIV 2.

N.E.C. Section 514-2

Above Ground Motor Fuel Pumps And Dispensers

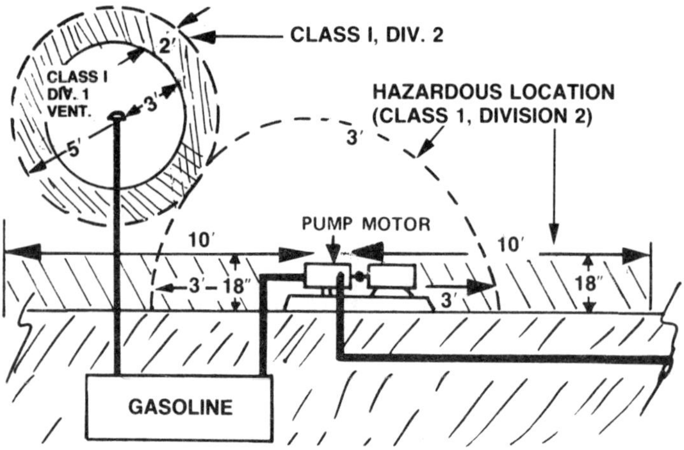

N.E.C. Section 514-2

Above Ground Motor Fuel Pumps And Dispensers

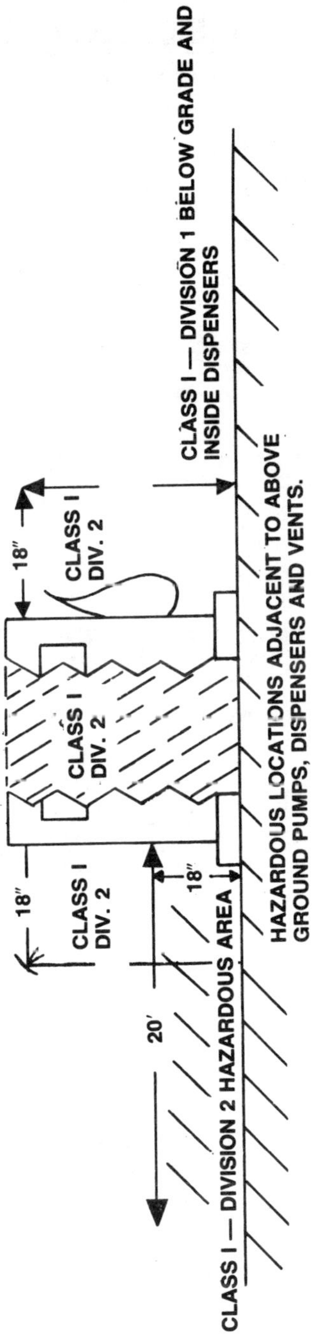

N.E.C. Section 514-5

Disconnecting Means

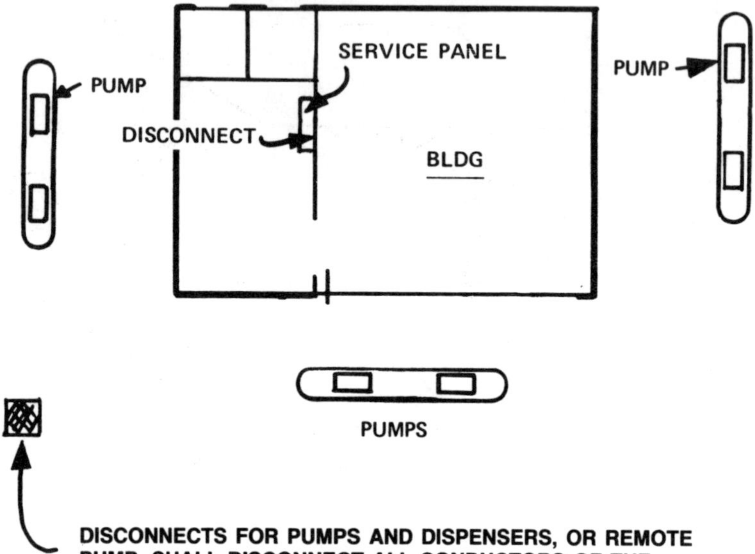

DISCONNECTS FOR PUMPS AND DISPENSERS, OR REMOTE PUMP. SHALL DISCONNECT ALL CONDUCTORS OF THE BRANCH CIRCUIT SIMULTANEOUSLY INCLUDING THE GROUNDED CONDUCTOR IF ANY.

N.E.C. Section 514-6

Sealing At Dispensing Apparatus

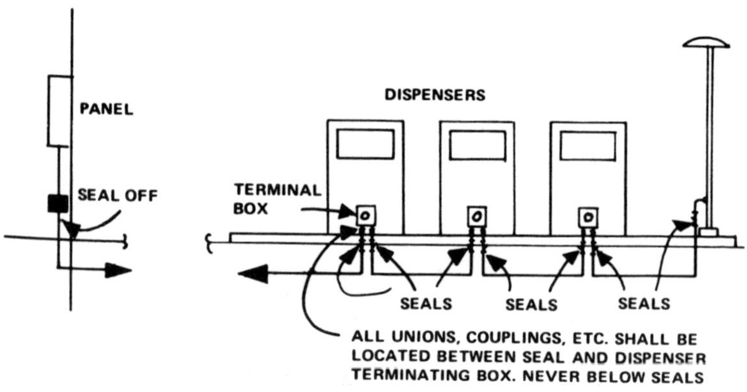

N.E.C. Table 515-2

Bulk Storage Plants

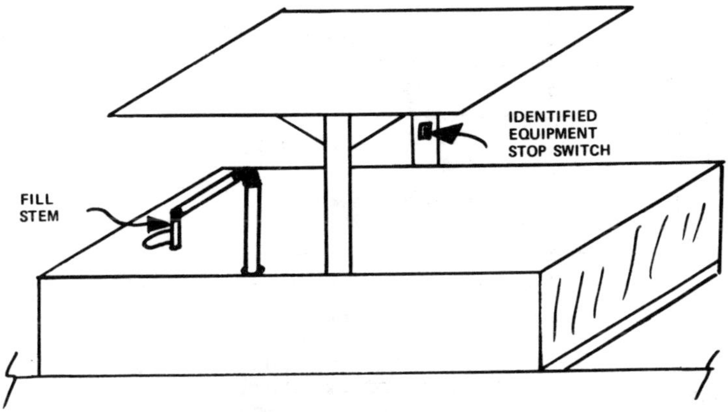

LOADING RACKS AND PLATFORMS
EXTENT OF HAZARDOUS AREA CLASS 1 DIV 2
10' HORIZONTALLY FROM FILL STEM AND 18" UPWARD
FROM GROUND LEVEL.

N.E.C. Table 515-2

Bulk Storage Plants

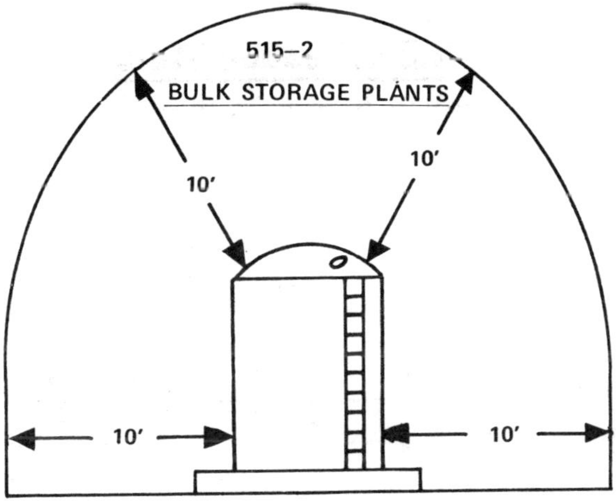

OUTDOOR TANKS ABOVE GROUND. ALL SPACE WITHIN
10' OF THE TANK IS A CLASS 1 DIV 2 HAZARDOUS LOCATION

N.E.C. Sections 516-2 (a) & (b) 2

Interlock For Spray Booth

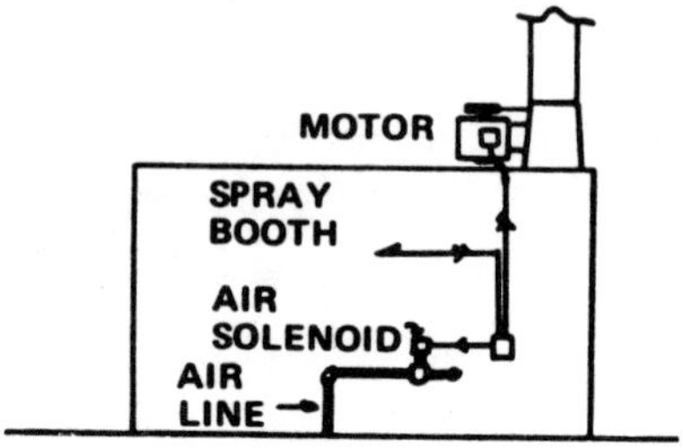

WHERE A SOLENOID VALVE IS INSTALLED IN THE AIR SUPPLY LINE TO THE BOOTH AND THE VALVE SHALL BE ELECTRICALLY INTERLOCKED WITH THE BOOTH EXHAUST SYSTEM TO PREVENT SPRAYING WHEN THE EXHAUST IS INOPERATIVE. ELECTRICAL CONNECTION BELOW

N.E.C. Section 516-3 (c)

Illumination Through Glass Panels

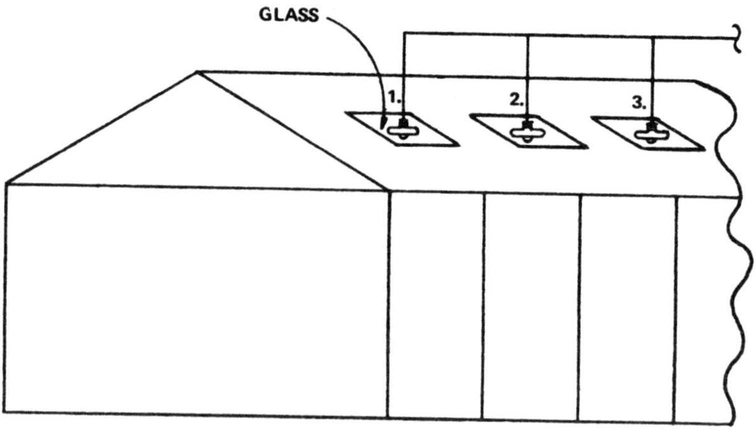

WHERE IN A HAZARDOUS LOCATION LIGHTING FIXTURES SHALL BE:

1. OF LISTED TYPE FOR ITS LOCATION.
2. SHIELDED FROM SPRAY AND RESIDUES; AND
3. FIXED IN PLACE AND ARRANGED TO BE LAMPED FROM OUTSIDE OF BOOTH.

N.E.C. Sections 516-3 & 516-4

Electrical Equipment In Spray Booths

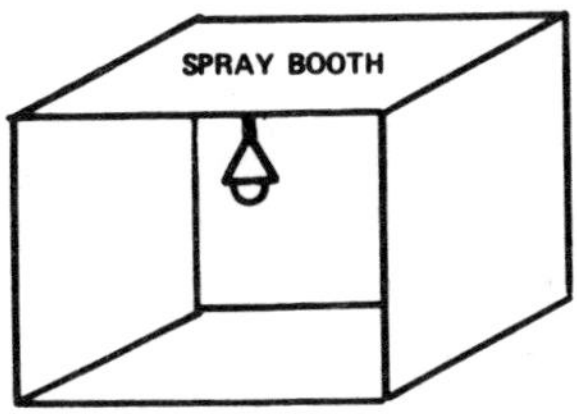

LIGHTS NOT APPROVED IN SPRAY BOOTHS.

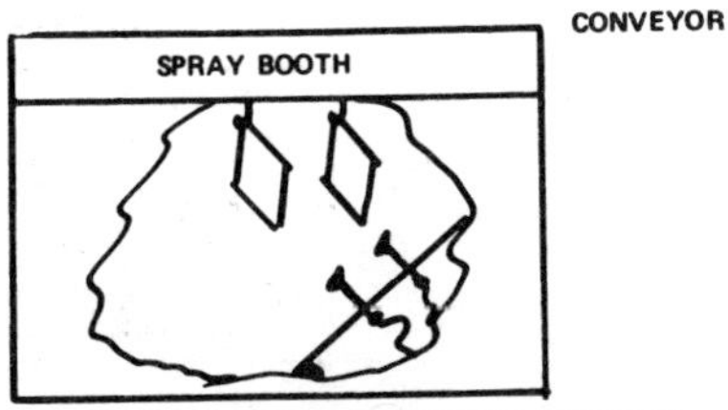

APPROVED ELECTRICAL EQUIPMENT LISTED FOR PURPOSE AND INSTALLED IN BOOTH IN ACCORDANCE WITH THE CONDITIONS OF LISTING.

N.E.C. Section 516-4

Electrostatic Equipment

ELECTROSTATIC SPRAYING AND DETEARING EQUIPMENT SHALL BE LISTED. INSTALLATION SHALL COMPLY WITH THE "CONDITIONS OF LISTING."

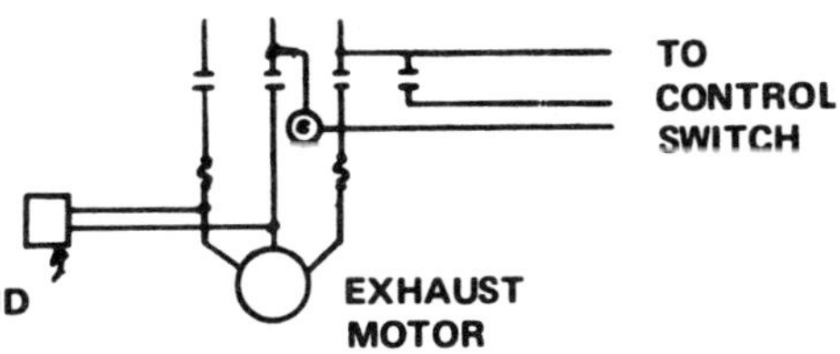

REDUCED AREA AROUND CLASS I DIV. 2 LOCATION WHERE AN INTERLOCK IS PROVIDED.

N.E.C. Section 516-7 (b)

Wiring Above Hazardous Locations

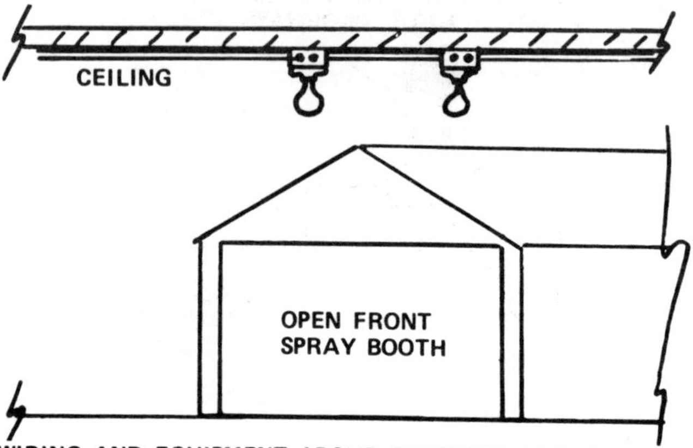

WIRING AND EQUIPMENT ABOVE FINISHING AREAS

EQUIPMENT WHICH MAY PRODUCE ARCS OR HOT METAL PARTICLES SHALL NOT BE INSTALLED ABOVE HAZARDOUS AREAS UNLESS TOTALLY ENCLOSED.

N.E.C. Sections 517-11, 13

Grounding In Patient — Care Areas

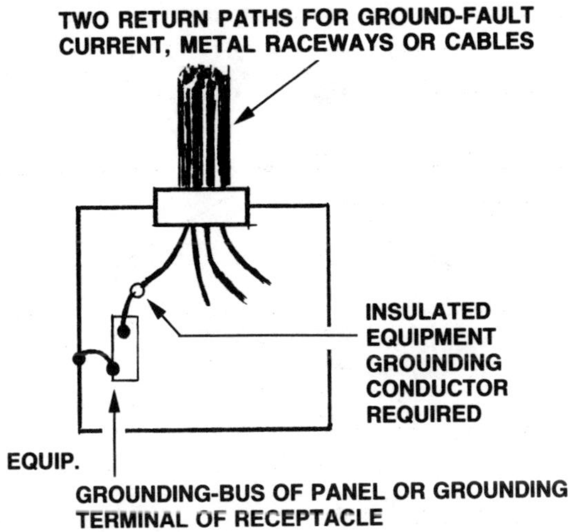

N.E.C. Section 517-19

Receptacle In Critical Care Areas

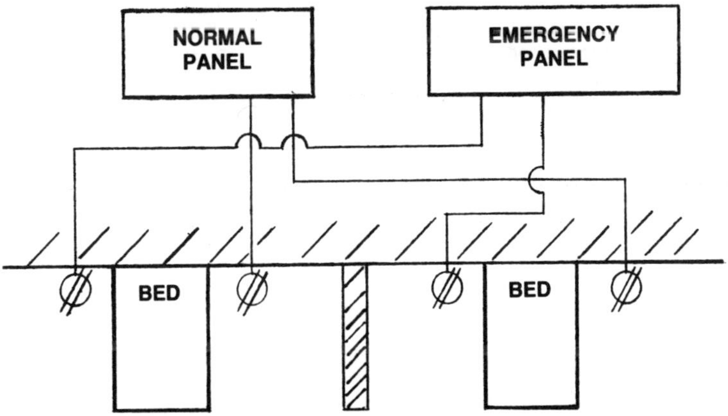

EACH BED MUST BE SUPPLIED BY AT LEAST:
- — ONE NORMAL BRANCH CIRCUIT
- — ONE EMERGENCY BRANCH CIRCUIT

EACH BED LOCATION SHALL BE PROVIDED WITH EITHER:
- — SIX RECEPTACLES: SINGLE, DUPLEX OR ANY COMBINATION

ALL THESE RECEPTACLES TO BE LISTED "HOSPITAL GRADE"

N.E.C. Section 517-19 (c)

Patient Care Areas Special Grounding Of Raceways

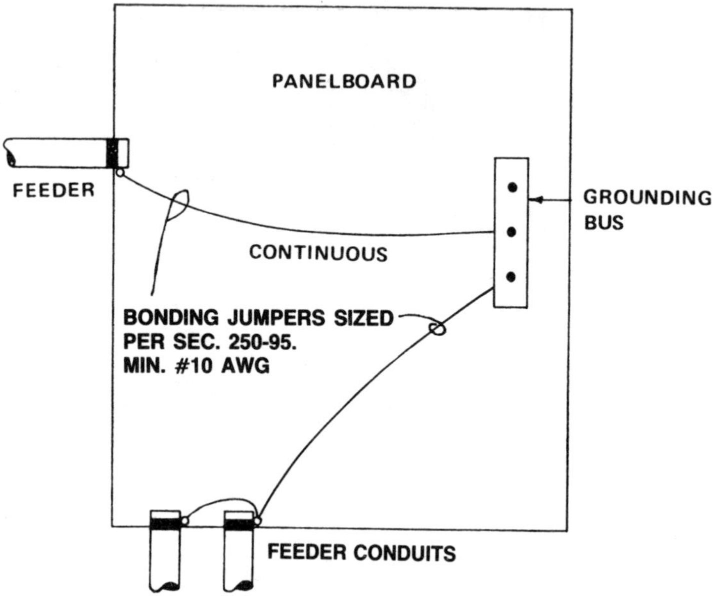

GROUNDING OF ALL FEEDER CONDUITS SHALL BE ASSURED BY MEANS OF GROUNDING BUSHINGS ON ALL TERMINATED CONDUITS & BONDED WITH AT LEAST NO. 10 STRANDED INSULATED COPPER CONDUCTORS TO THE GROUNDING BUS.

N.E.C. Sections 517-60, 61

Anesthetizing Locations

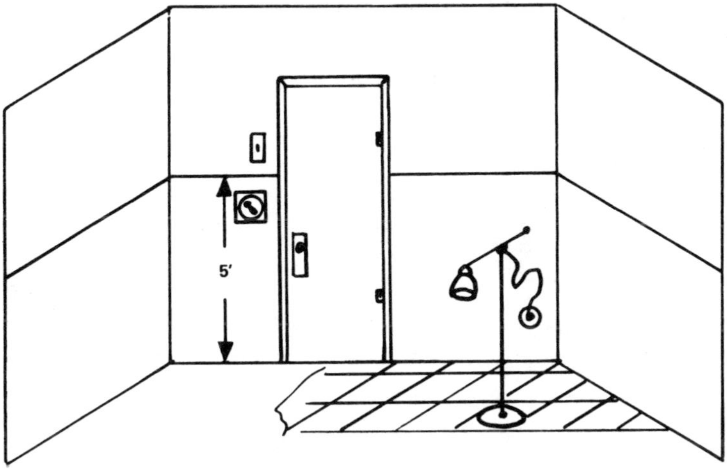

ANESTHETIZING LOCATIONS SHALL COMPLY WITH PROVISIONS OF SECTION 517-61, ETC. WHETHER FIXED OR PORTABLE (EXCEPT CERTAIN EQUIPMENT OPERATING AT 10 VOLTS OR LESS)

N.E.C. Section 517-60

Combustible Anesthetics

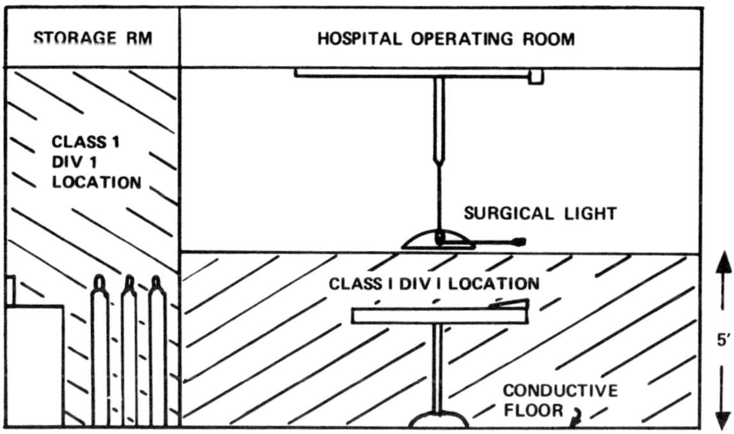

N.E.C. Section 517-61

Circuits In Anesthetizing Locations

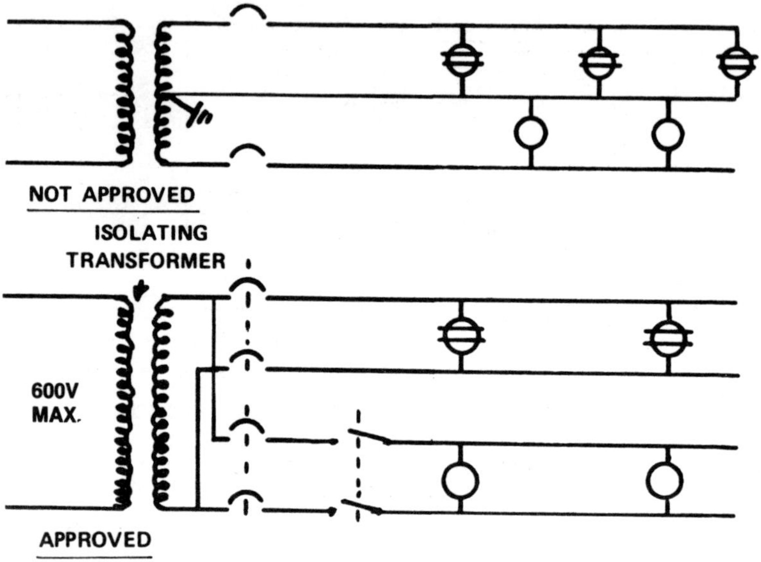

CIRCUITS IN ANESTHETIZING LOCATIONS (EXCEPT FOR LIGHTS OVER 8' HIGH) SHALL BE ISOLATED FROM OTHER CIRCUITS IN THE BUILDING AND FROM GROUND.

N.E.C. Section 600-2

Sign Disconnects

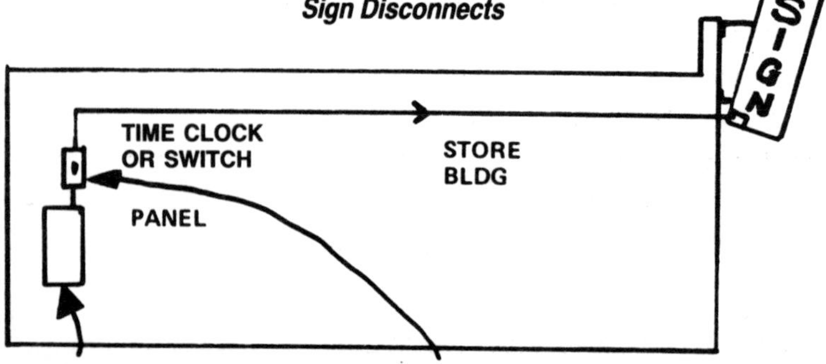

CIRCUIT BKR. OR SWITCH REQUIRED TO CONTROL SIGN.
NOTE: U.L. STANDARDS REQUIRE SEPARATE ACCESSIBLE COMPARTMENTS FOR FUSEHOLDERS OR SWITCHES WITHIN THE BODY OF THE SIGN.

N.E.C. Section 600-3

Use Of Signs As Junction Boxes

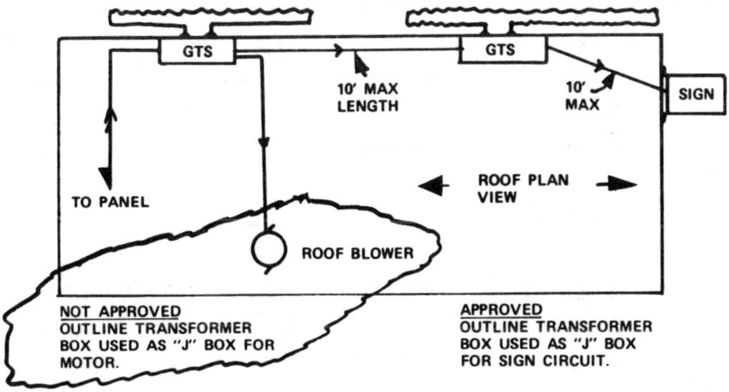

N.E.C. Section 600-6 (a)

Load Of Branch Circuit

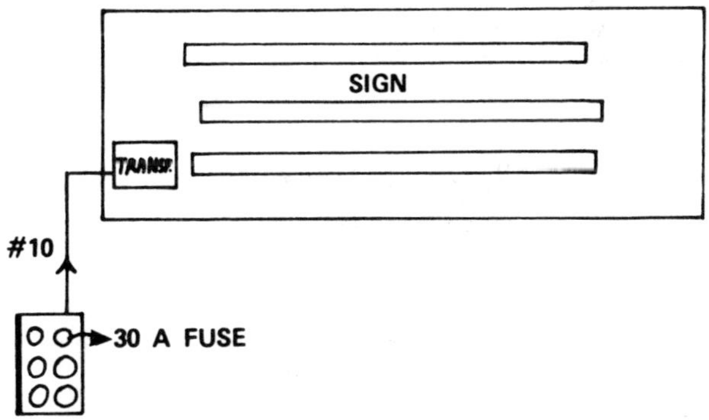

APPROVED 30 A. CIRCUITS CONTAINING ELECTRIC-DISCHARGE LIGHTING TRANSFORMERS EXCLUSIVELY.

N.E.C. Section 600-6 (a)

Branch Circuit Ratings For Signs

NOT APPROVED.

BRANCH CIRCUIT OF OVER 20 AMP RATING SUPPLYING TRANSFORMERS FOR SIGNS OR OUTLINE LIGHTS.

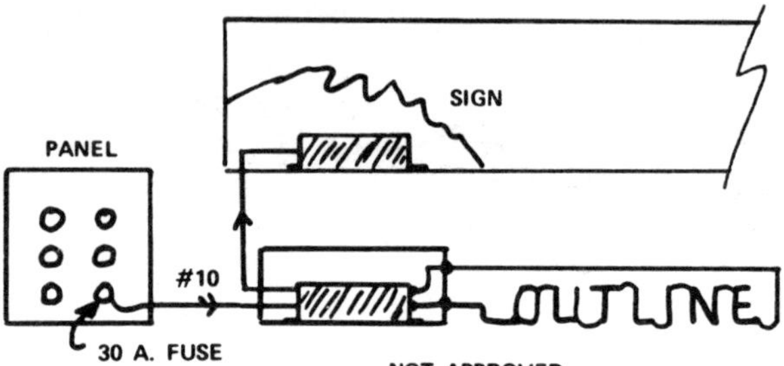

NOT APPROVED

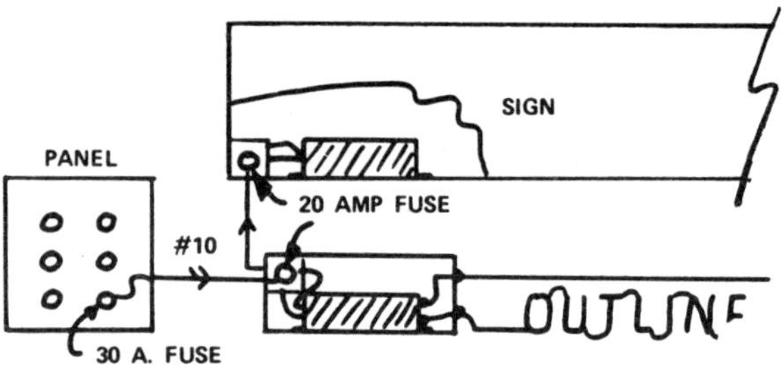

APPROVED

CIRCUIT PROTECTED AT OVER 20 AMPS SUPPLYING TRANSFORMERS FOR SIGN AND OUTLINE LIGHTING WITH 20 AMP FUSES IN SIGN AND OUTLINE ENCLOSURES.

N.E.C. Section 600-6 (b) & (c)

Sign Circuit

A MINIMUM OF ONE BRANCH CIRCUIT COMPUTED AT NOT LESS THAN 1200 VA IS REQUIRED TO BE INSTALLED *SOLELY* FOR THE SUPPLY OF EXTERIOR SIGNS OR OUTLINE LIGHTING FOR EACH GROUND FLOOR COMMERCIAL BUILDING OR STORE OCCUPANCY.

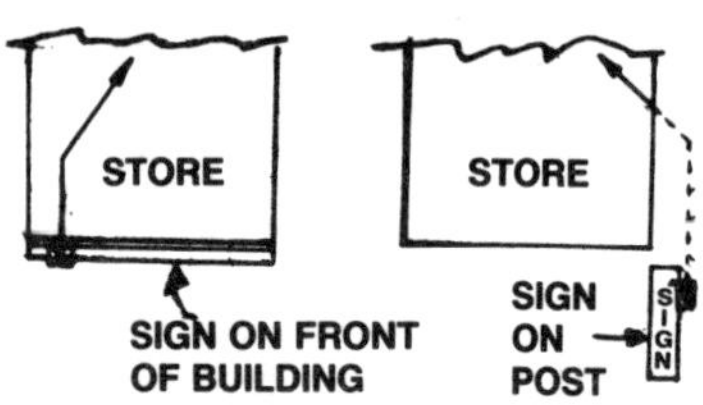

N.E.C. Section 600-36 (b), (c)

Interior Skeleton Signs

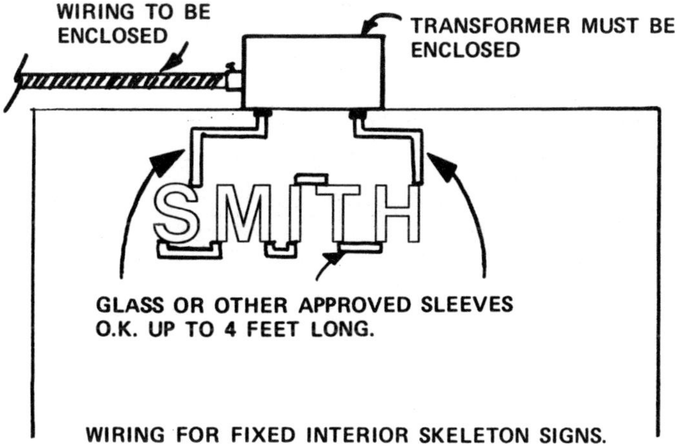

N.E.C. Section 600-37 (a), (c), (d)

Portable Signs

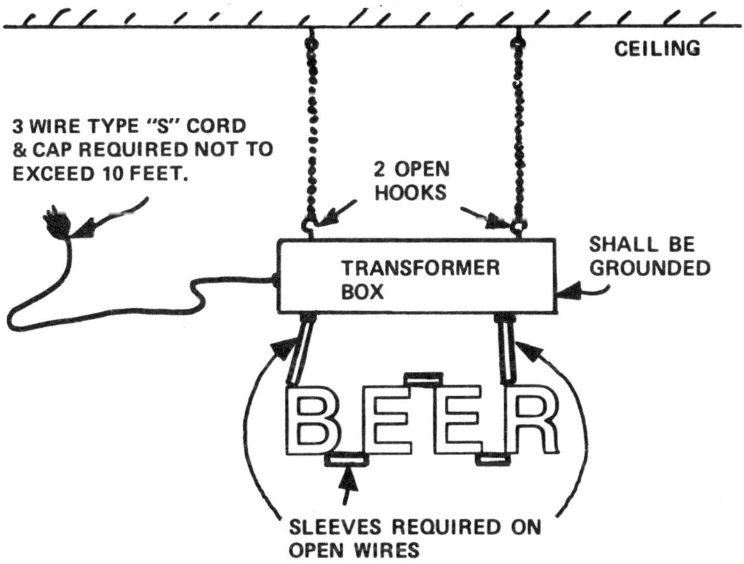

PORTABLE GAS TUBE SIGNS ARE FOR INTERIOR USE ONLY

N.E.C. Section 610-21 (a)

Crane Contact Conductors

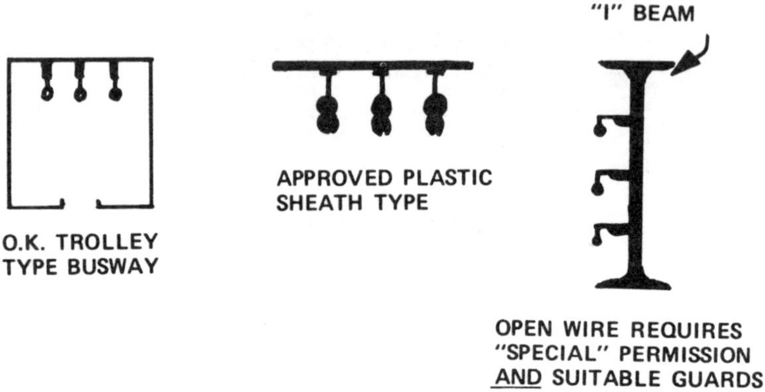

N.E.C. Section 610-21 (c)

Supports For Crane Contact Conductors

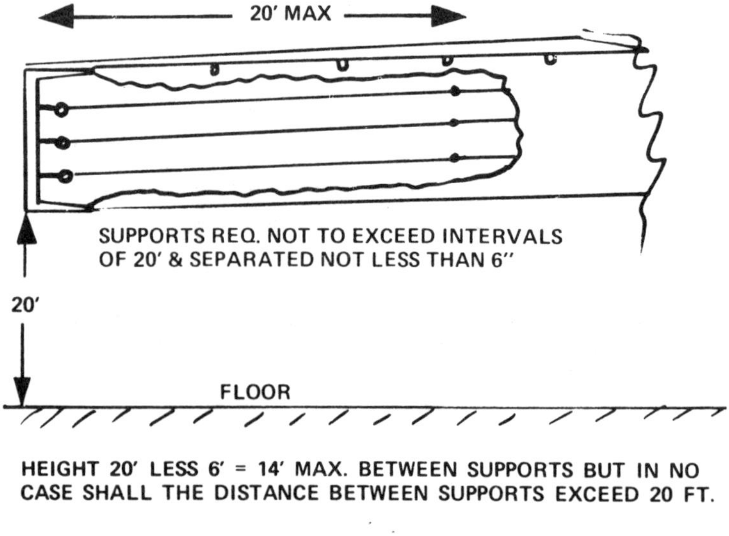

HEIGHT 20' LESS 6' = 14' MAX. BETWEEN SUPPORTS BUT IN NO CASE SHALL THE DISTANCE BETWEEN SUPPORTS EXCEED 20 FT.

N.E.C. Section 610-42

Cab Operated Crane Provisions

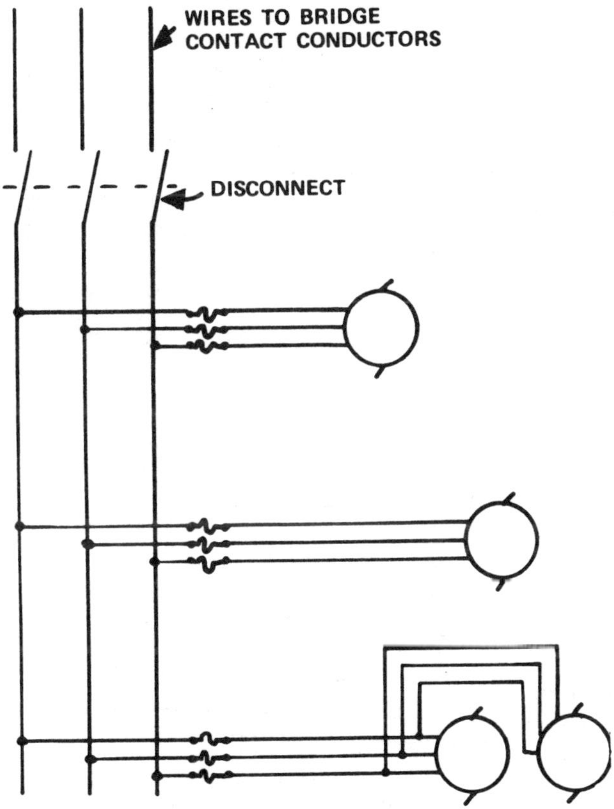

CAB OPERATED CRANE

O.K. MOTORS PROTECTED BY FUSES OF SIZE PERMITTED IN TABLE 430-152. ONE SET OF FUSES O.K. FOR TWO MOTORS USED FOR SINGLE OPERATION.

N.E.C. Section 620-22

Separate Circuits For Elevators

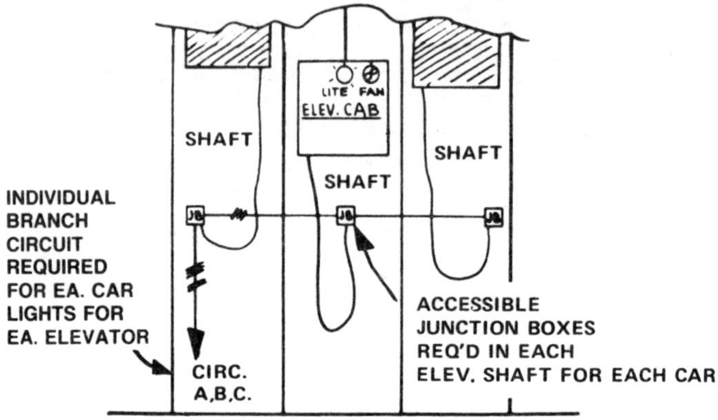

N.E.C. Section 620-37

Foreign Wiring In Hoist Ways

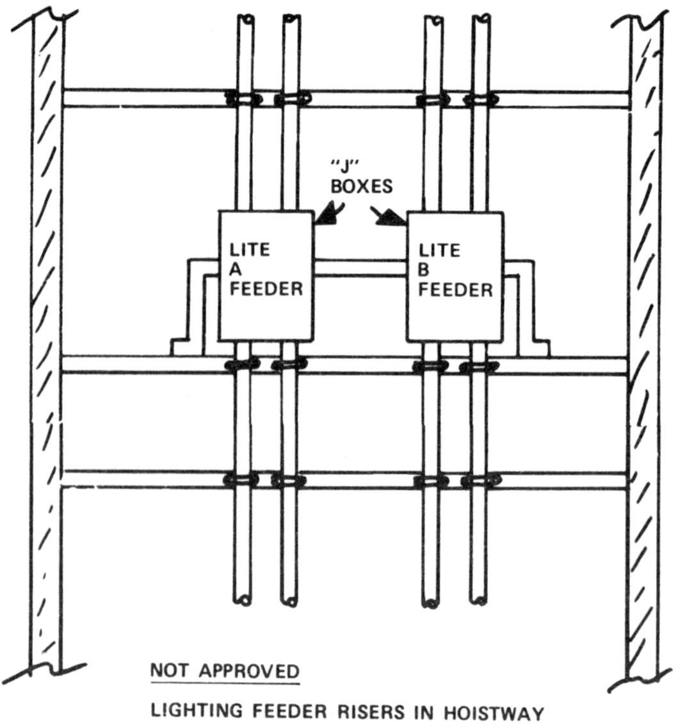

N.E.C. Section 630-11 (a)

Rating Of Conductors For Transformer Arc Welders

CALCULATING THE RATED CURRENT OF TRANSFORMER ARC WELDERS TO DETERMINE THE AMPACITY OF THE SUPPLY CONDUCTORS.

FOR AN INDIVIDUAL WELDER:

MULTIPLY THE PRIMARY NAMEPLATE CURRENT OF THE WELDER BY THE FOLLOWING FACTOR: (BASED UPON DUTY CYCLE OR TIME RATING OF THE WELDER)

RATED PERCENT DUTY CYCLE OF WELDER.	MULTIPLING FACTOR
20 OR LESS	0.45
30	0.55
40	0.63
50	0.71
60	0.78
70	0.84
80	0.89
90	0.95
100	1.00
WELDER HAVING TIME RATING OF 1 HR.	.75

FOR GROUPED WELDERS:

THE AMPACITY OF THE SUPPLY CONDUCTORS MAY BE LESS THAN THE SUM OF THE CALCULATED RATED CURRENTS OF ALL WELDERS SUPPLIED SEE SEC. 630–11(b)

EXAMPLE

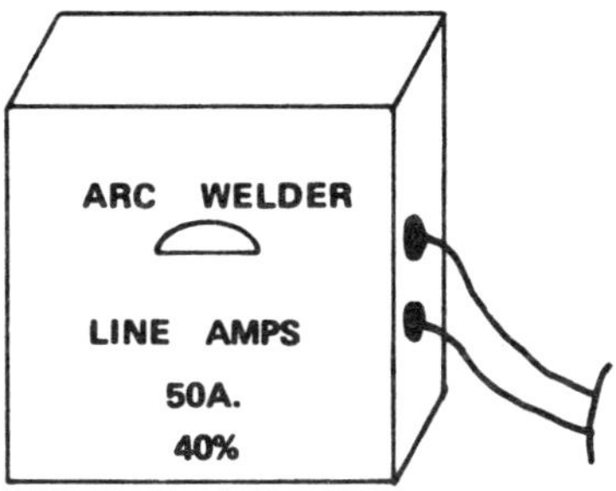

LINE AMPS X FACTOR = RATED AMPS.

50A X 0.63 = 31.5 AMPS

N.E.C. Sections 630-12

Overcurrent Protection For Transformer Arc Welders

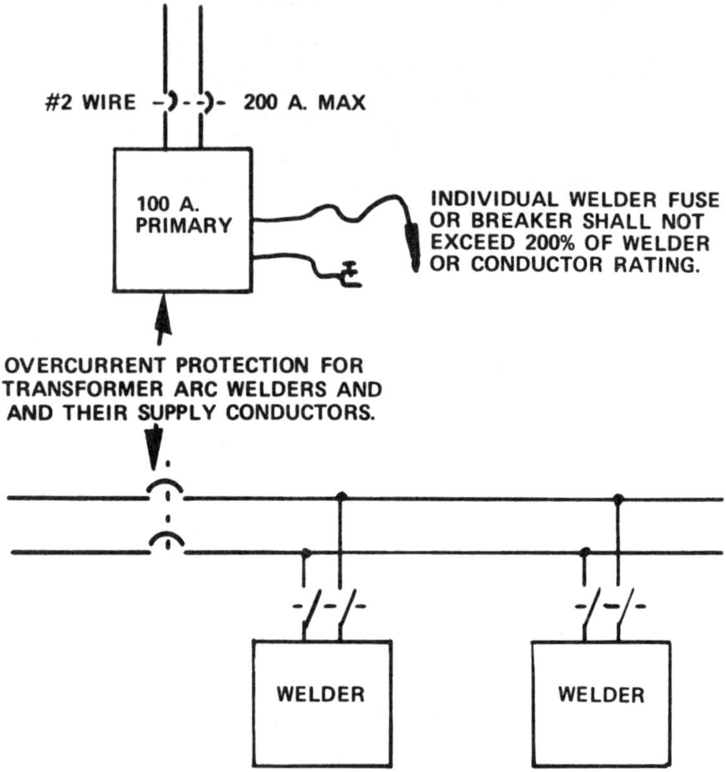

GROUP OF WELDERS
FUSE OR BREAKER SHALL NOT EXCEED 200% OF RATING OF SUPPLY CONDUCTORS OR SMALLEST WELDER.

N.E.C. Section 630-31

Conductors For Resistance Welders

SUPPLY CONDUCTORS FOR RESISTANCE WELDERS

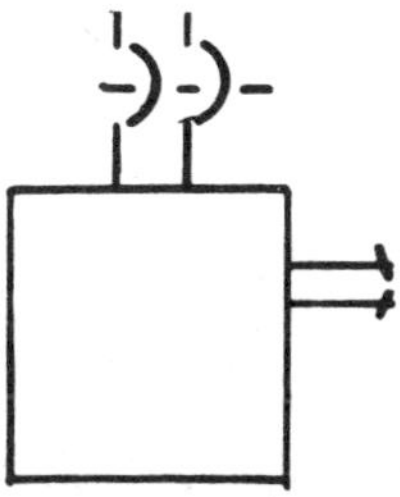

INDIVIDUAL WELDER SPECIFIC OPERATION. PRIMARY CURRENT AND DUTY CYCLE REMAIN UNCHANGED-RATING OF SUPPLY CONDUCTORS DETERMINED FROM ACTUAL CURRENT, USING MULTIPLIER.

INDIVIDUAL WELDER GENERAL DUTY. RATING OF SUPPLY CONDUCTORS SHALL BE RATED NOT LESS THAN 70% OF RATED PRIMARY OF WELDER FOR AUTOMATIC OR SEAM WELDERS OR 50% FOR MANUAL OPERATED NON-AUTOMATIC WELDERS.

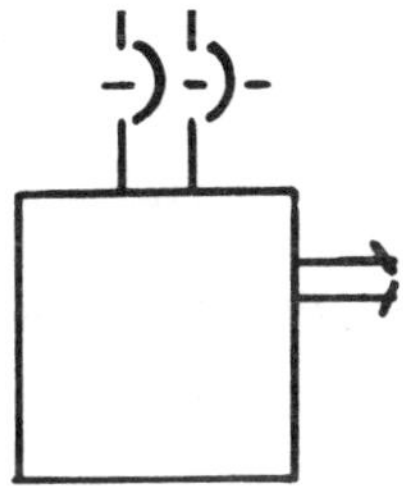

N.E.C. Section 630-32

Overcurrent Protection For Resistance Welders

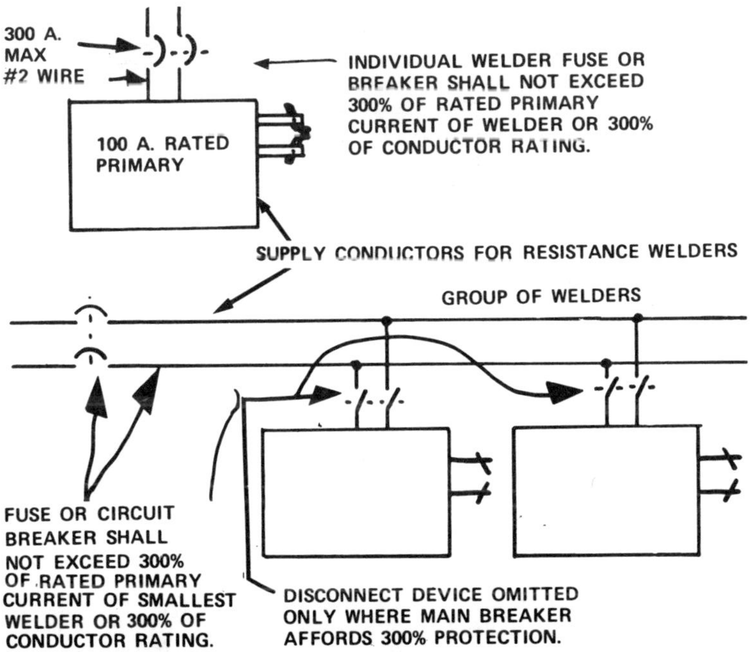

N.E.C. Section 645-10

Data Processing Room Disconnecting Means

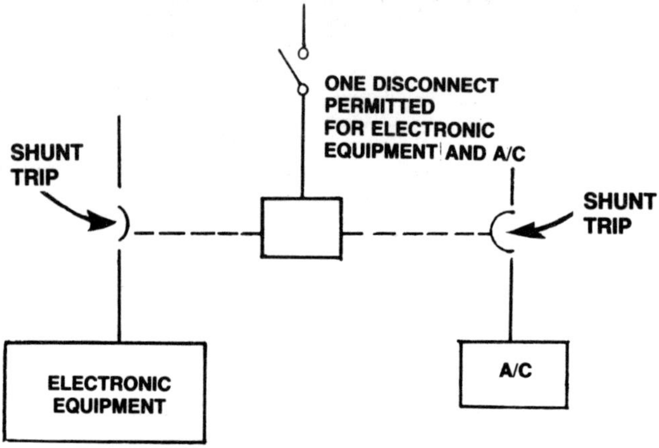

N.E.C. Section 670-3

Machine Tool Nameplate Data

— *SINGLE MOTOR* —

DIXIE MACHINE COMPANY
MFD. IN POINT BARROW, ALASKA
VOLTS 240 — PHASE 3 — 60 CYCLE
AMPERES 62 H.P. 25 AND SHORT-CIRCUIT
INTERRUPTING CAPACITY OF THE MACHINE
OVERCURRENT PROTECTIVE DEVICES.

THIS MARKING IS ADEQUATE FOR A SINGLE MOTOR MACHINE. THE NAMEPLATE SHALL BE IN PLAIN SIGHT AND COULD BE ON THE MOTOR WHERE IT IS OBVIOUS THAT IT IS A SINGLE MOTOR MACHINE.

— *MULTI MOTOR* —

DIXIE MACHINE COMPANY
MFD. IN POINT BARROW, ALASKA
VOLTS 480 – PHASE 3 – 60 CYCLE

MAIN DRIVE MOTOR	HP 10	AMPS 14
COOLANT MOTOR	HP 1/6	AMPS .4
VERTICAL DRIVE MOTOR	HP 2	AMPS 6.5
HORIZONTAL DRIVE MOTOR	HP 1	AMPS 3.5
FEED MOTOR HP	¾	AMPS 2.8
MAXIMUM AMPS	20.9	

THIS MARKING IS FOR A MULTI-MOTOR MACHINE. EACH MOTOR IS GIVEN AND THE MAXIMUM AMPACITY IS 20.9 (ASSUMING MAIN, COOLANT AND VERTICAL DRIVE MOTORS TO BE OPERATIVE AT ONE TIME).

N.E.C. Sections 680-20 (b), 680-22 & 680-25 (b)

Pool Equipment Bonding & Grounding

1. An insulated 16 copper conductor provided integral with the cord for the pool light.

2. An insulated 12 copper conductor shall be installed between the pool junction box and the panel.

   EXCEPTION:

   Where two or more pool lights are supplied by the same circuit they are permitted to be terminated at the grounding terminals in the pool light junction box with one 12 grounding conductor, grounded at the panel.

3. A 12 copper conductor shall be provided in each conduit to ground all other electrical equipment such as recirculating pump motors, time clocks, etc.

4. Separate conductors are required for grounding of pool lights and for bonding of pool equipment.

   EXAMPLE:

   Where PVC is used from the pool junction box, one unbroken insulated 12 conductor is required for the ground and an additional conductor for the bonding of the equipment.

N.E.C. Sections 680-20 (b) 1, 680-22 & 680-25 (d)

Pool Equipment Bonding And Grounding

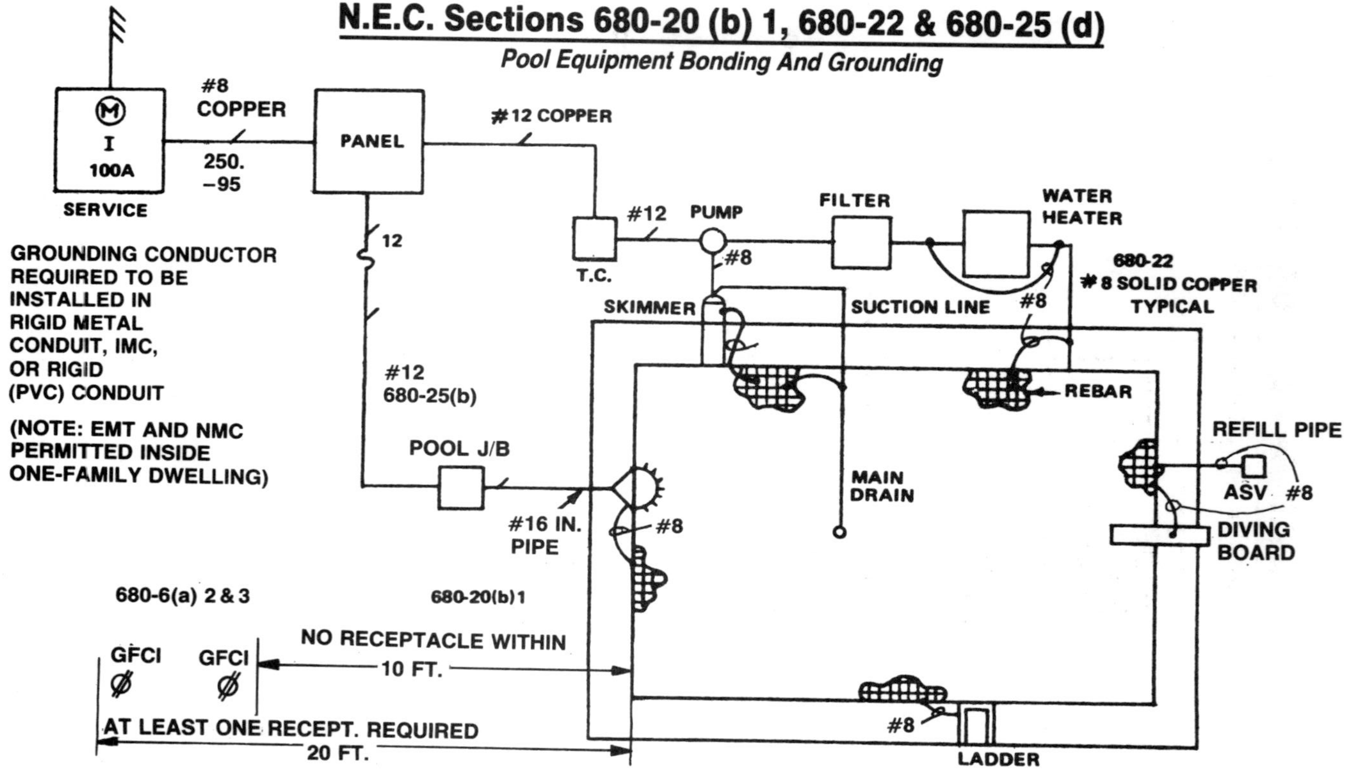

N.E.C. Section 700-6 — 17

Emergency Lighting Transfer Switch

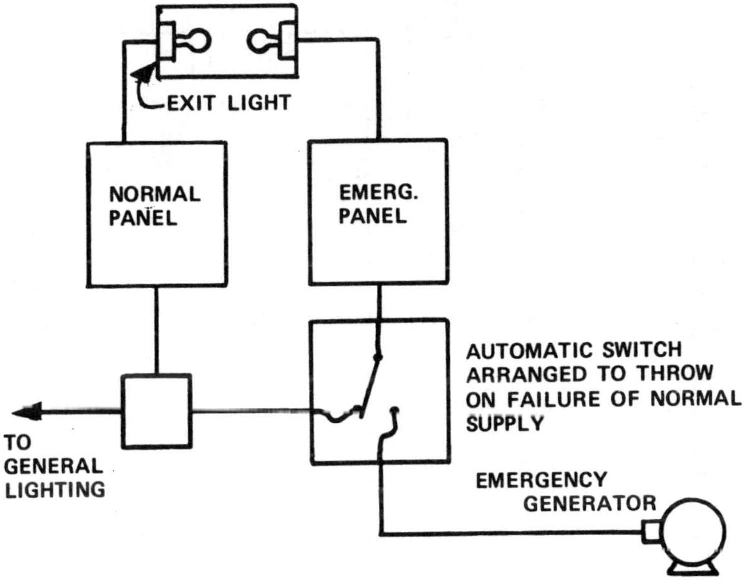

N.E.C. Section 700-12 (f)

Unit Equipment

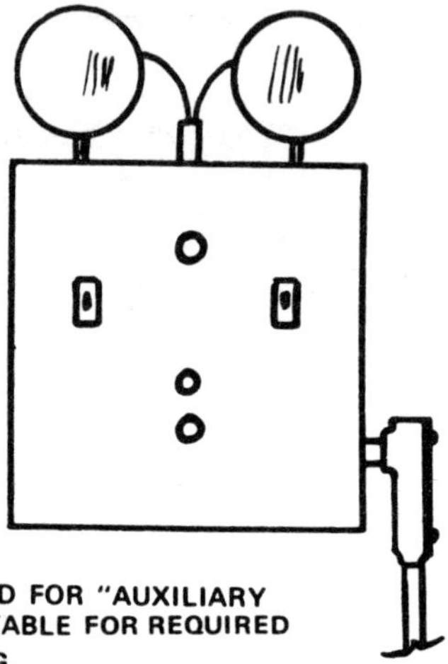

NOTE:
EQUIPMENT APPROVED FOR "AUXILIARY USE" <u>IS NOT</u> ACCEPTABLE FOR REQUIRED EMERGENCY LIGHTING.

N.E.C. Sections 700-15, 700-16 & 700-17

Automatic Transfer Switch

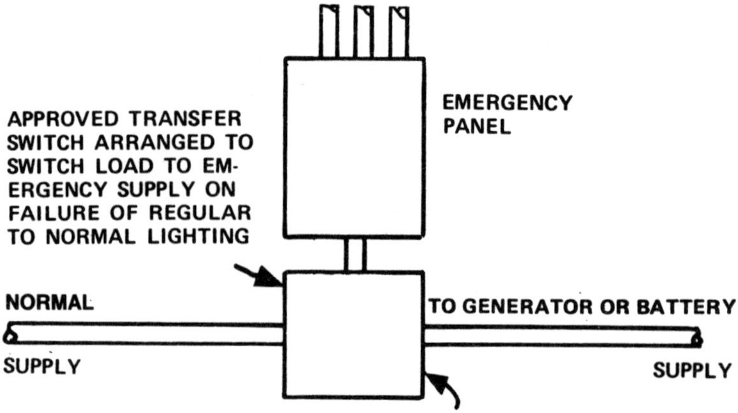

EMERGENCY LIGHTING MAY BE SUPPLIED FROM GENERAL LIGHTING SERVICE WHERE AN AUTOMATIC TRANSFER SWITCH IS USED.

N.E.C. Section 700-16

Egress Lighting And Exit Sign Requirements

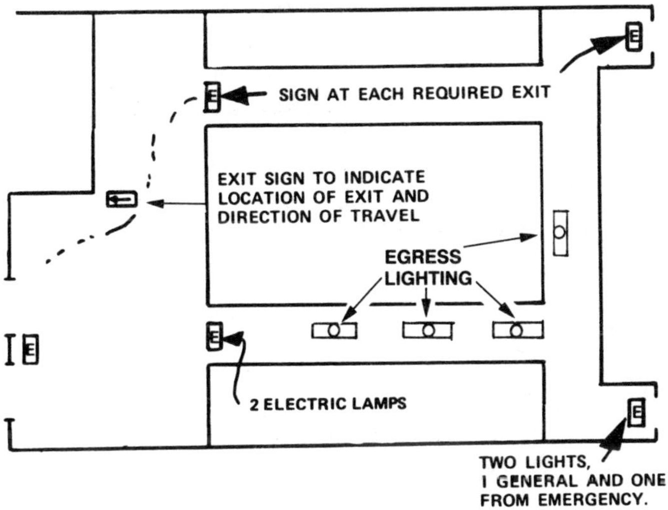

N.E.C. Section 700-16

Required Emergency Lighting

(THESE DRAWING ARE AN INTERPRETATION OF SECTION 700-16. CHECK WITH LOCAL JURISDICTION FOR LOCAL INTERPRETATION)

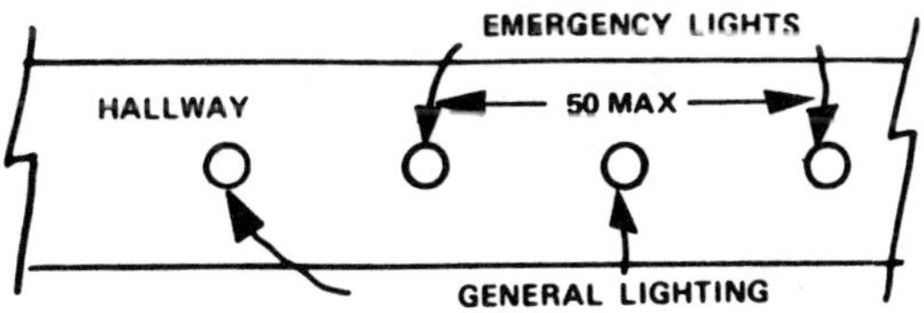

HALLWAYS, LOBBIES, OR SIMILAR EXIT PASSAGEWAY (IN OCCUPANCIES REQUIRING EMERGENCY SYSTEMS)

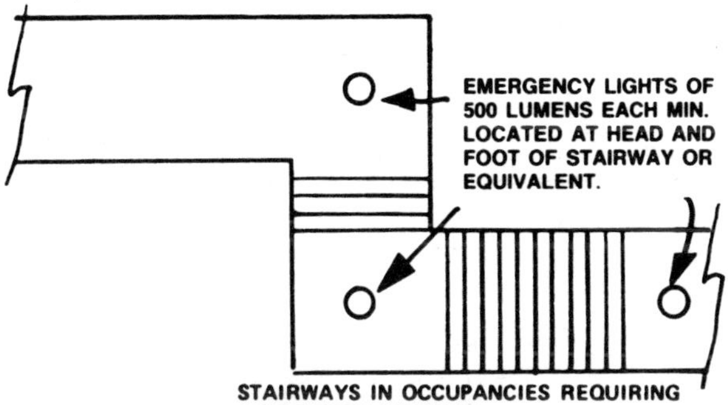

STAIRWAYS IN OCCUPANCIES REQUIRING EMERGENCY SYSTEMS.

N.E.C. Section 700-17

Independent Wiring System Required

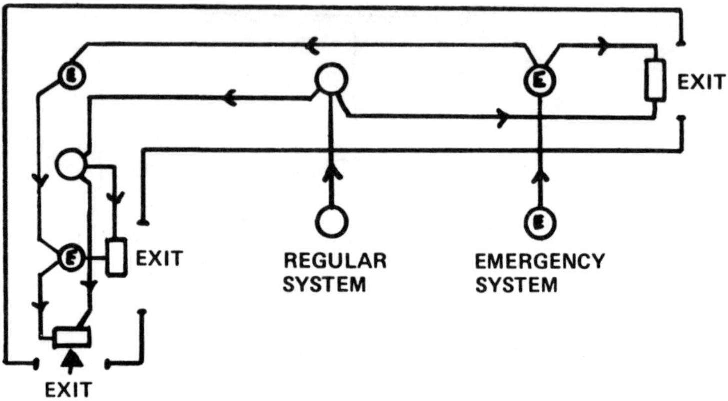

EMERGENCY SYSTEM SHALL NOT ENTER THE SAME RACEWAY, BOX, OR CABINET WITH OTHER SYSTEMS. (EXCEPT EXIT BOXES, TRANSFER SWITCHES, ETC.)

N.E.C. Section 700-9 (a)

Wiring-Emergency System

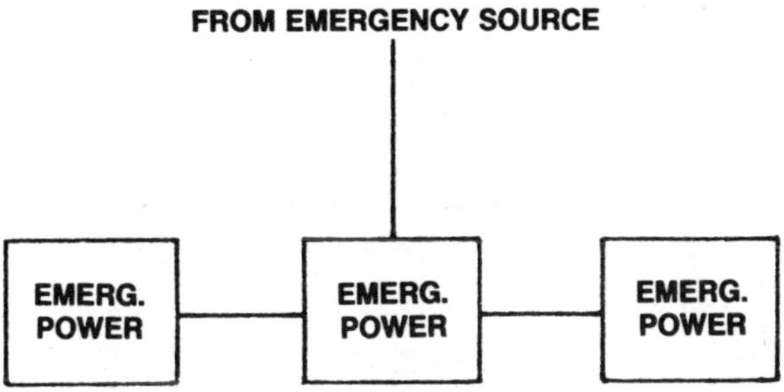

ALL BOXES AND ENCLOSURES FOR EMERGENCY CIRCUITS TO BE IDENTIFIED

N.E.C. Section 700-12 (f) Exception

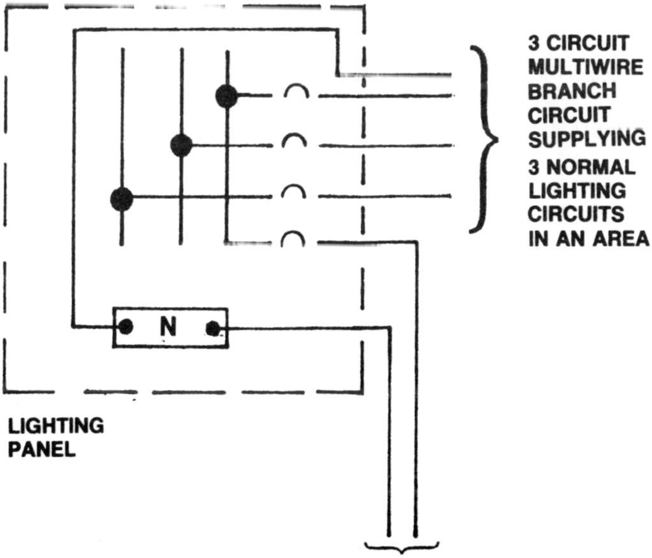

THEN A SEPARATE BRANCH CIRCUIT MAY SUPPLY EMERGENCY LIGHTING UNITS FROM THE SAME PANEL

N.E.C. Section 710-3 (b)

High Voltage Conductors

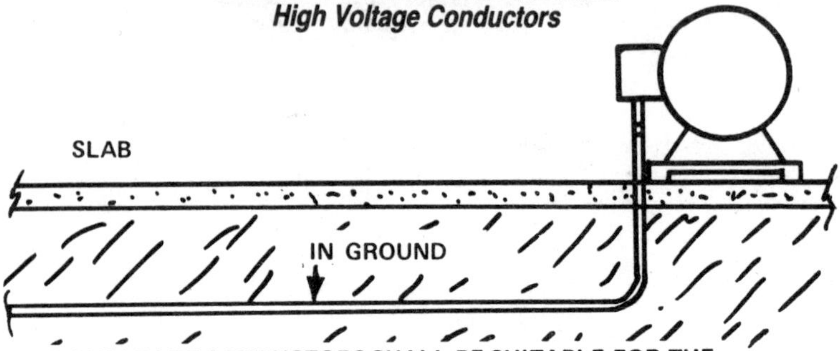

HIGH VOLTAGE CONDUCTORS SHALL BE SUITABLE FOR THE CONDITIONS OF USE. WIRE IN CONDUIT IN THE EARTH SHALL BE SUITABLE FOR THE VOLTAGE AND WET LOCATION.

N.E.C. Section 710 — 3(a) & (b)

General Requirements — Circuits Over 600 Volts, Nominal

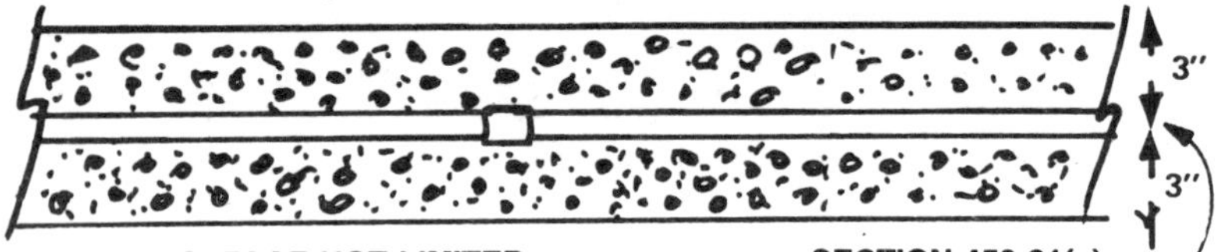

VOLTAGE NOT LIMITED.

SECTION 450-21(c)

ENCASEMENT IN CONCRETE ONLY REQUIRED FOR NONSHIELDED CABLES.

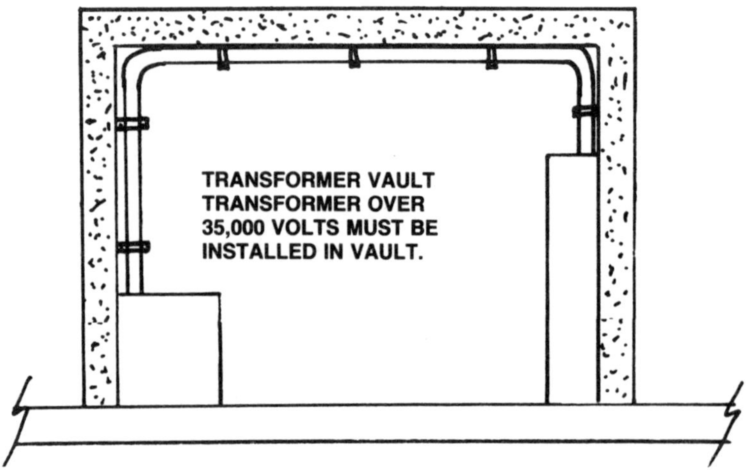

TRANSFORMER IN VAULTS COMPLYING WITH 450-42 VOLTAGE IS LIMITED TO 35,000 VOLTS.

N.E.C. Section 710-21

Interrupting Capacity

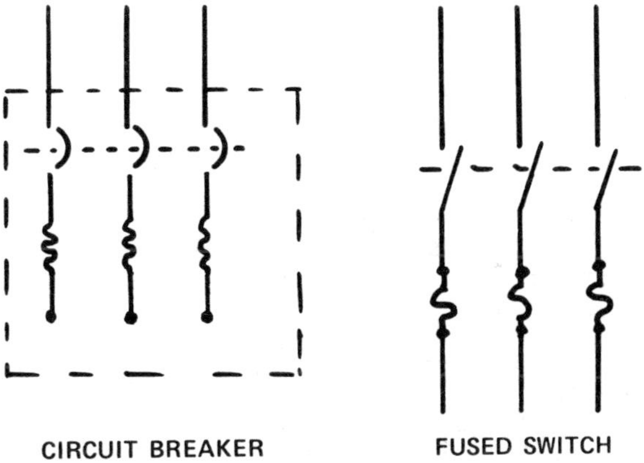

FUSES OR CIRCUIT BREAKERS SHALL HAVE AN INTERRUPTING RATING AT LEAST EQUAL TO THE MAXIMUM SHORT CIRCUIT POSSIBLE IN THE CIRCUIT THEY PROTECT.

N.E.C. Section 710-21 (b) & (e)

Fused Switches

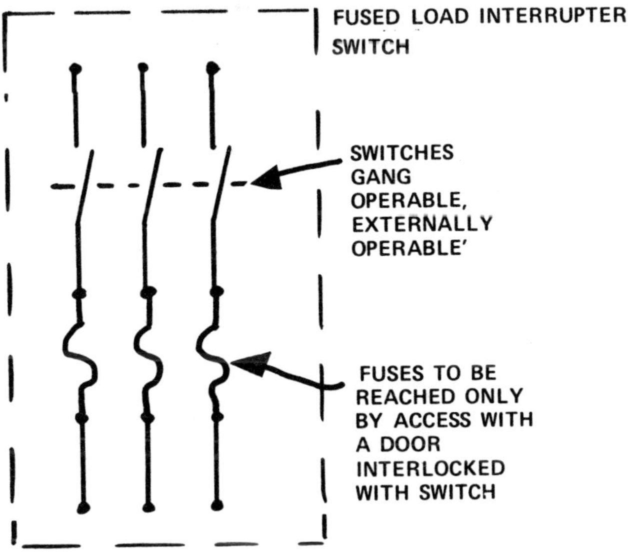

SWITCHES REQUIRED TO HAVE INTERRUPTING RATING AT LEAST EQUAL TO THE LOAD SUPPLIED.

N.E.C. Section 710-21 (d)

Oil Filled Cutouts

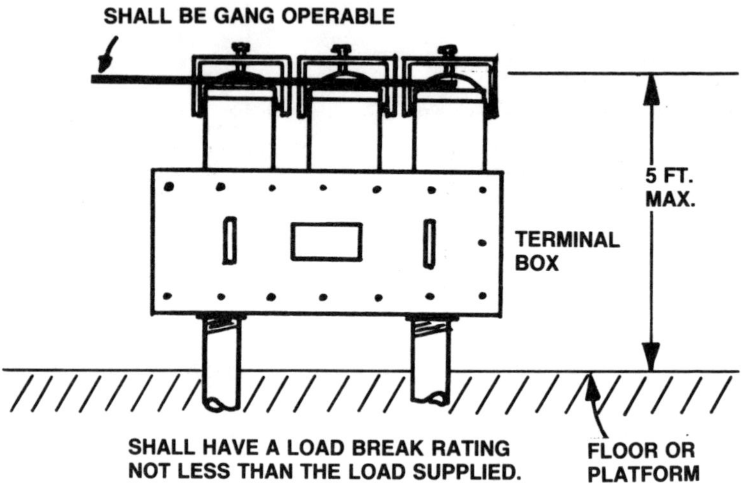

N.E.C. Section 710-22

Isolating Switches

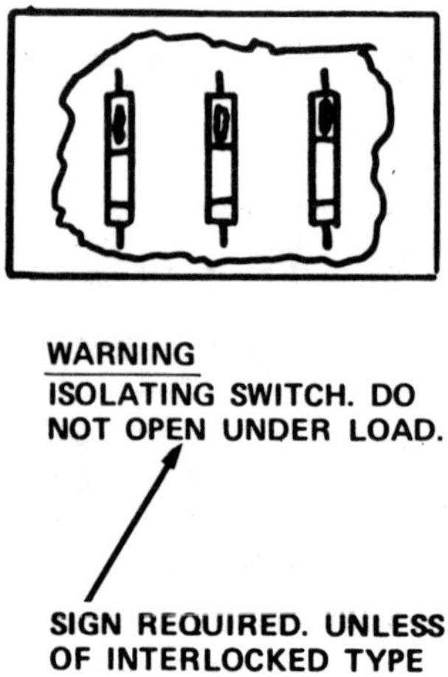

N.E.C. Section 725-4

Circuits For Safety Control Devices

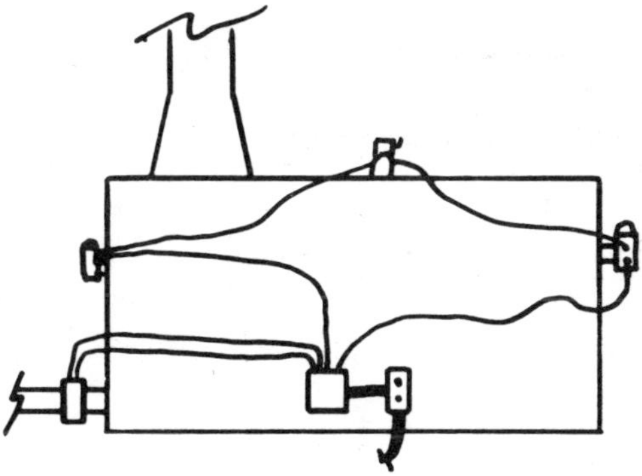

NOT APPROVED

SAFETY CONTROL WIRING ON BOILER WITH OPEN WIRE NOT APPROVED REGARDLESS OF VOLTAGE.

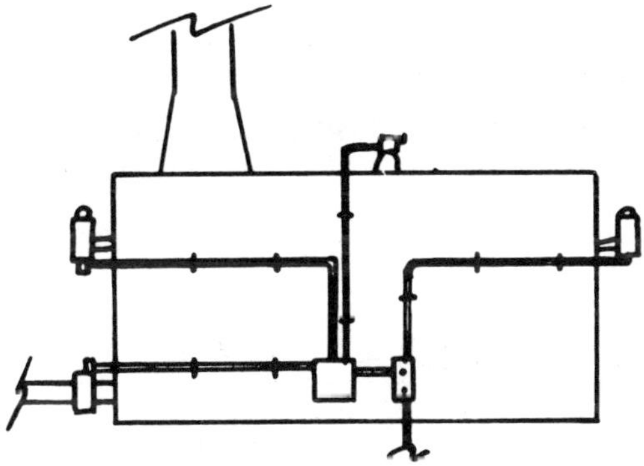

APPROVED

SAFETY CONTROLS WIRED AS A CLASS I SYSTEM.

N.E.C. Section 725-11 – 12

Class 1 Transformer

TRANFORMERS OR OTHER EQUIPMENT SUPPLYING CLASS 1 SYSTEMS SHALL BE IDENTIFIED FOR THE PURPOSE OR FOR GENERAL USE.

PROTECTION OF TRANSFORMERS FOR CLASS 1 SYSTEMS:

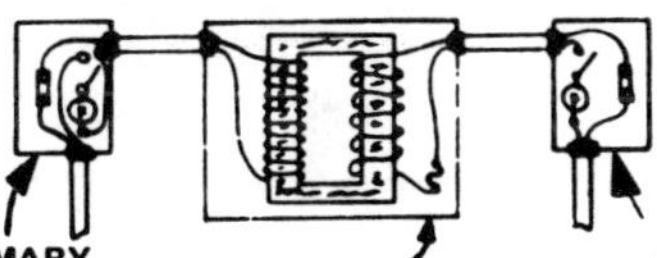

PRIMARY OVERCURRENT DEVICE OR

BUILT–IN INTEGRAL OVERCURRENT DEVICE OR

SECONDARY OVERCURRENT DEVICE

N.E.C. Section 725-14

Class 1 Wiring Method

THE INSTALLATION OF CLASS I WIRING AND EQUIPMENT MUST BE INSTALLED IN APPROVED RACEWAYS AND ENCLOSURES COMPLYING WITH THE GENERAL WIRING METHODS OF CHAPTER 3 OF THE CODE.

OPEN WIRING NOT APPROVED

EXCEPTION:

DOES NOT APPLY WHERE OTHER ARTICLES OF THIS CODE SPECIFICALLY PERMIT OR REQUIRE OTHER METHODS FOR REMOTE CONTROL OR SIGNAL CIRCUITS.

EXAMPLES:

ART. 500 HAZARDOUS LOCATIONS

ART. 620 ELEVATORS-DUMBWAITERS-AND ESCALATORS

ETC.

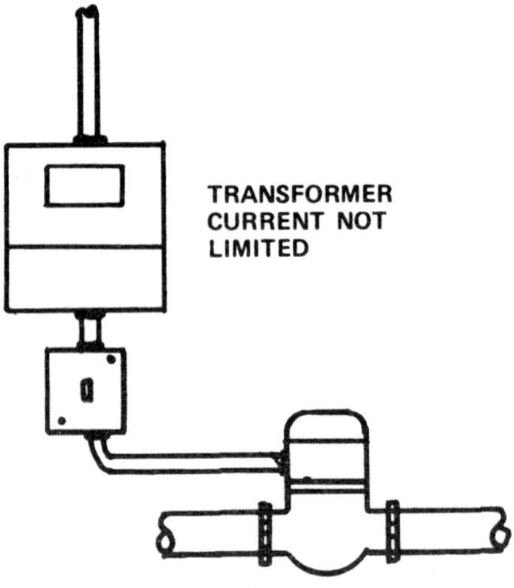

WIRING FOR CLASS I CIRCUITS SHALL BE INSTALLED IN ACCORDANCE WITH THE REQUIREMENTS OF CHAPTER 3 OF THIS CODE.

N.E.C. Sections 725-14 & 725-16 (a)

Conductor Sizes For Class I Systems

SHALL NOT BE SMALLER THAN NO. 14
EXCEPT:

1–NO. 18 OR NO. 16 CONDUCTORS MAY BE USED PROVIDED THEY SUPPLY LOADS THAT DO NOT EXCEED THE AMPACITIES GIVEN IN SECTION 402-5 AND INSTALLED IN RACEWAY OR LISTED CABLE.

2–FLEXIBLE CORDS MAY BE USED WHERE INSTALLED AS REQUIRED IN ART. 400 (FLEXIBLE CORDS).

3–CABLED CONDUCTORS SUITABLE FOR THE APPLICATION MAY BE USED.

N.E.C. Section 725-16 (b)

Insulation Of Conductors For Class I Systems

ALL CONDUCTORS TO HAVE 600V INSULATION.

1–INDIVIDUAL CONDUCTORS OF NO. 14 OR LARGER SHALL COMPLY WITH ART. 310.

2–INDIVIDUAL CONDUCTORS NO. 18 OR 16 SHALL BE AS LISTED IN SECTION 725-16b.

OR

3–CONDUCTORS APPROVED FOR THE PURPOSE

N.E.C. Section 725-31, 725-34

Class 2 System Current And Voltage Limitations

CLASS 2 LIMITED POWER SOURCES

A LISTED CLASS 2 TRANSFORMER

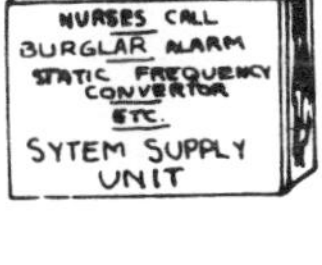

ENERGY LIMITED POWER SOURCE LISTED FOR THE PURPOSE AND USED IN ACCORDANCE WITH THE CONDITIONS OF THE LISTING.

N.E.C. Sections 725-38 (b) & 725-40

Class 2 And Class 3 Conductors

CLASS 2 AND CLASS 3 CONDUCTORS SHALL BE APPROVED FOR THE PURPOSE OR SUITABLE FOR THE CONDITIONS INVOLVED.

CONDUCTORS CAN BE SINGLE OR CABLED WITH RUBBER, THERMOPLASTIC OR SIMILAR TYPE OF INSULATION.

OPEN WIRING MUST BE A MULTI-CONDUCTOR SHIELDED, ARMORED, OR JACKETED TYPE CABLE, AND APPROVED FOR USE AS A LOW ENERGY CIRCUIT CABLE.

N.E.C. Section 725-38 (a) (2)

Wiring For Class 2 System

NOT APPROVED

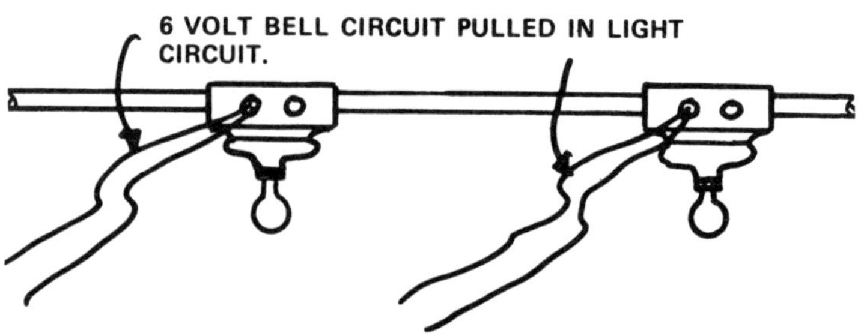

WIRING FOR CLASS 2 SYSTEMS SHALL NOT BE RUN IN ENCLOSURES WITH LIGHTING AND POWER WIRING.
EXCEPT:
WHERE LOOMED IN TRANSFORMER BOXES OR WHERE RUN AS CLASS I SYSTEM.

APPENDIX

SERVICE ENTRANCE GROUNDING AND BONDING DIAGRAMS

Prepared by
The Southwestern Section
International Association of Electrical Inspectors
Grounding Committee
Clive LeVoir, Chairman

These diagrams are intended to illustrate typical grounding and bonding requirements but are not intended to show the only acceptable grounding and bonding methods.

Based on the 1993 National Electrical Code.

COMBINATION METER AND SERVICE SECTION FOR OVERHEAD DWELLING SERVICE

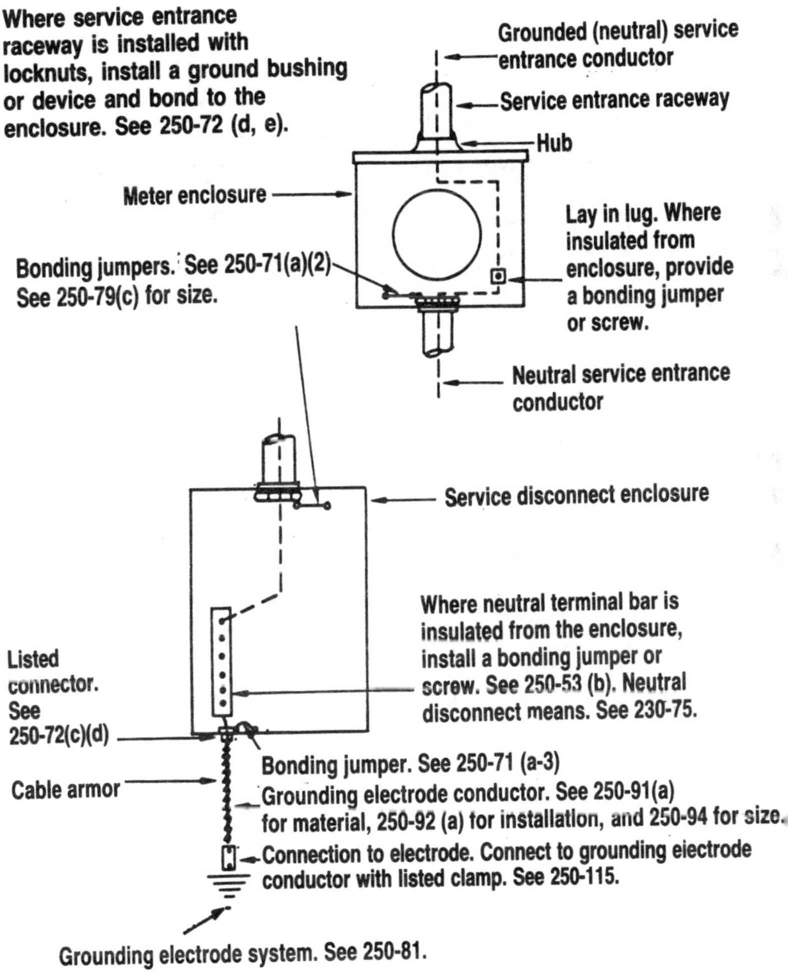

COMBINATION METER AND SERVICE SECTION FOR OVERHEAD DWELLING SERVICE

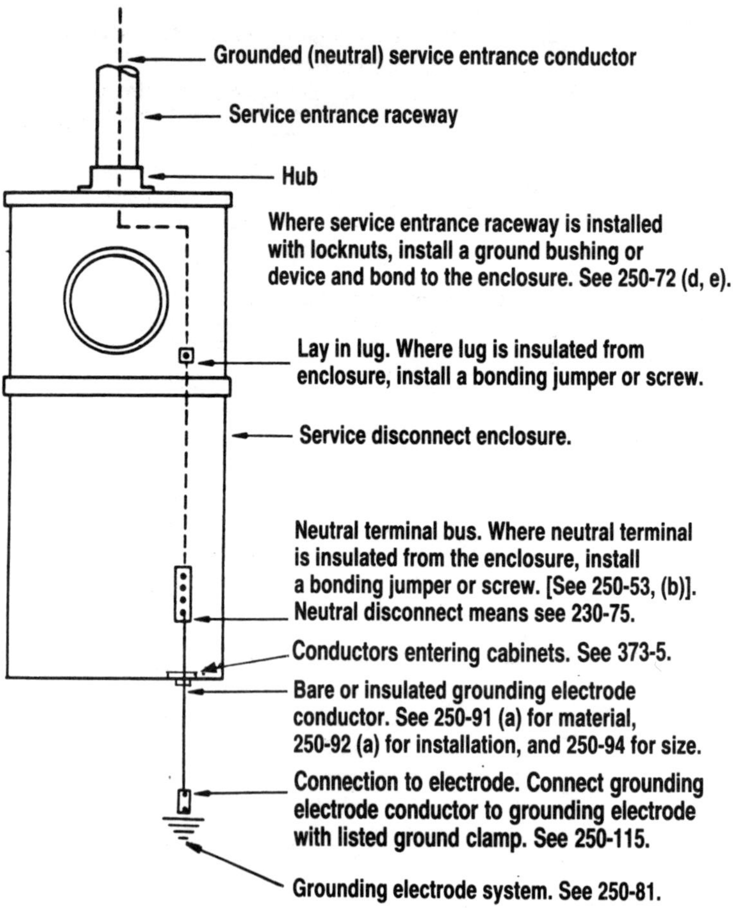

MULTI-FAMILY DWELLING SERVICE SIX OR LESS SUBDIVISIONS

Where service entrance racewaway is installed with locknuts, install a ground bushing or device and bond to the enclosure. See 250-72 (d, e).

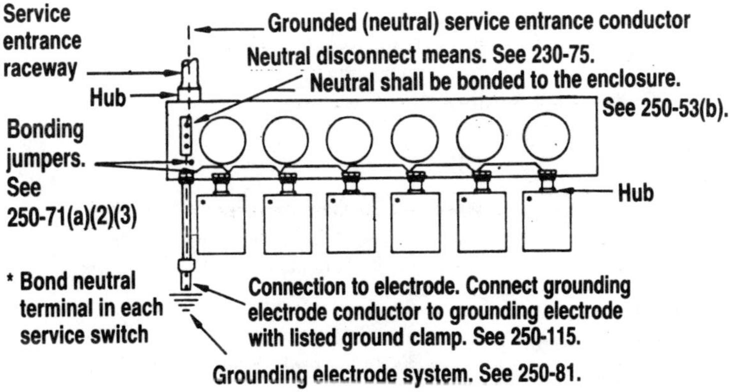

NOTE: *Metallically isolated water piping systems. See 250-80.*

MULTI-FAMILY DWELLING SERVICE MORE THAN SIX SUBDIVISIONS

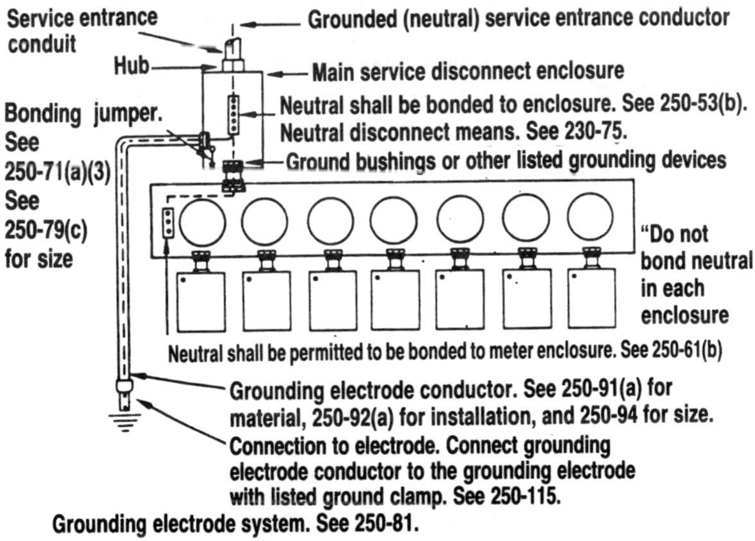

NOTE: *Metallically isolated water piping systems. See 250-80.*

METER AND SERVICE WITH CURRENT TRANSFORMER ENCLOSURE OVERHEAD SUPPLY

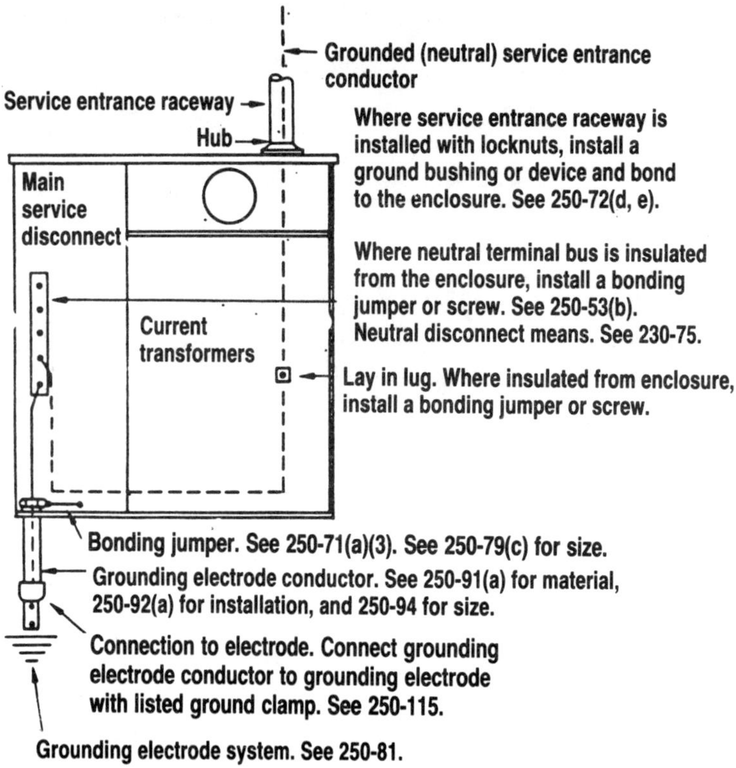

SINGLE METER COMMERCIAL SERVICE WITH A NEUTRAL

(WHERE A MAIN DISCONNECTING MEANS IS REQUIRED)

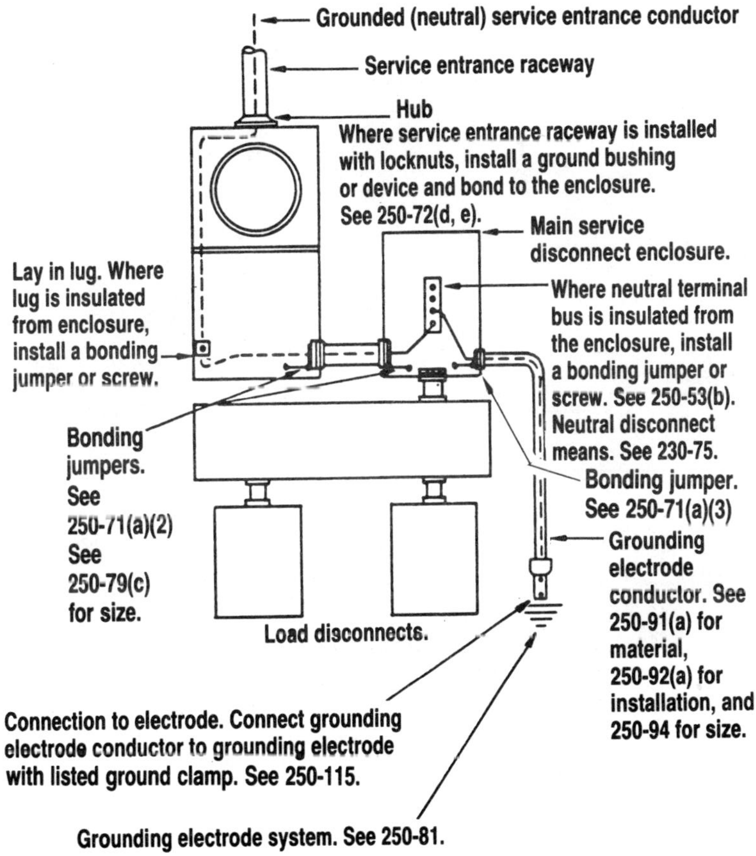

THREE-PHASE, THREE-WIRE DELTA SERVICE WITH ADDED GROUNDING CONDUCTOR*

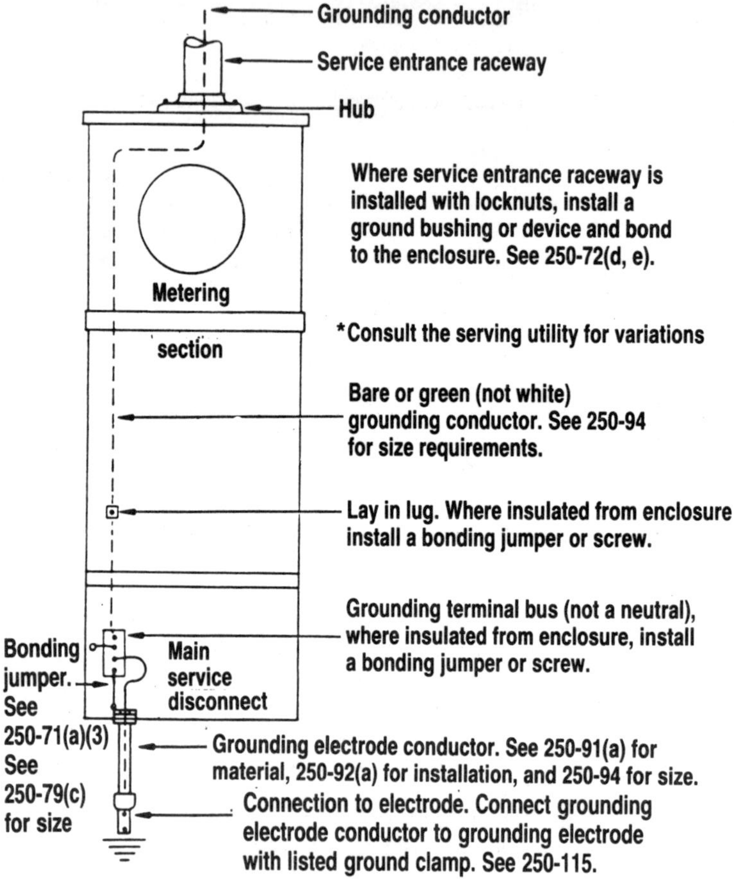

Grounding electrode system. See 250-81.

COMMERCIAL SERVICE OVERHEAD SUPPLY SIX SUBDIVISIONS OR LESS

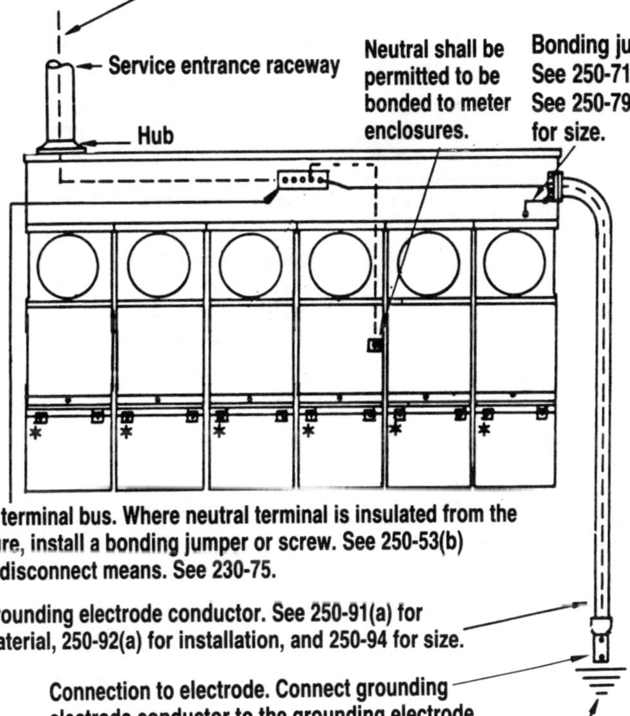

NOTE: ***Metallically isolated water piping systems. See 250-80.***

COMMERCIAL SERVICE OVERHEAD SUPPLY MORE THAN SIX SUBDIVISIONS OF SERVICE

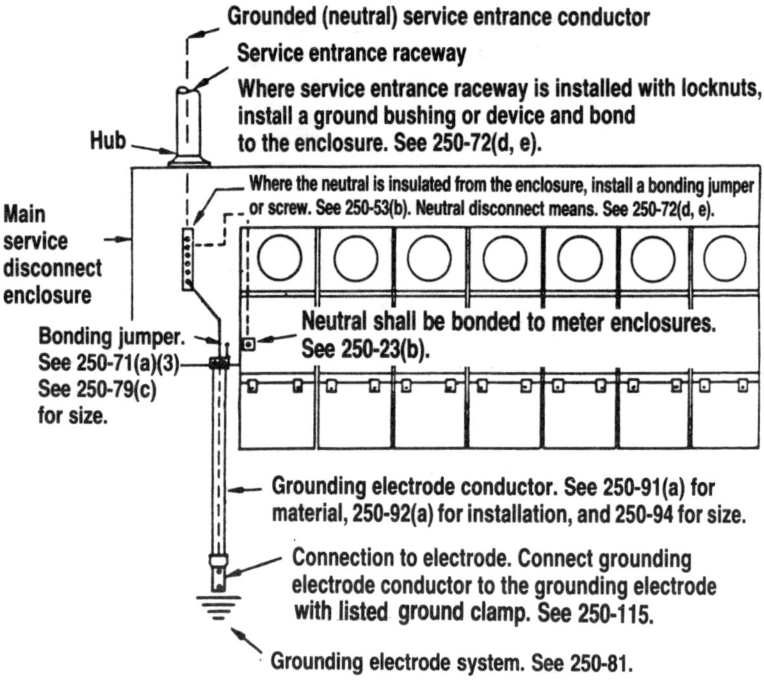

NOTE: *Metallically isolated water piping systems. See 250-80.*

COMBINATION METER AND SERVICE FOR UNDERGROUND DWELLING SERVICE

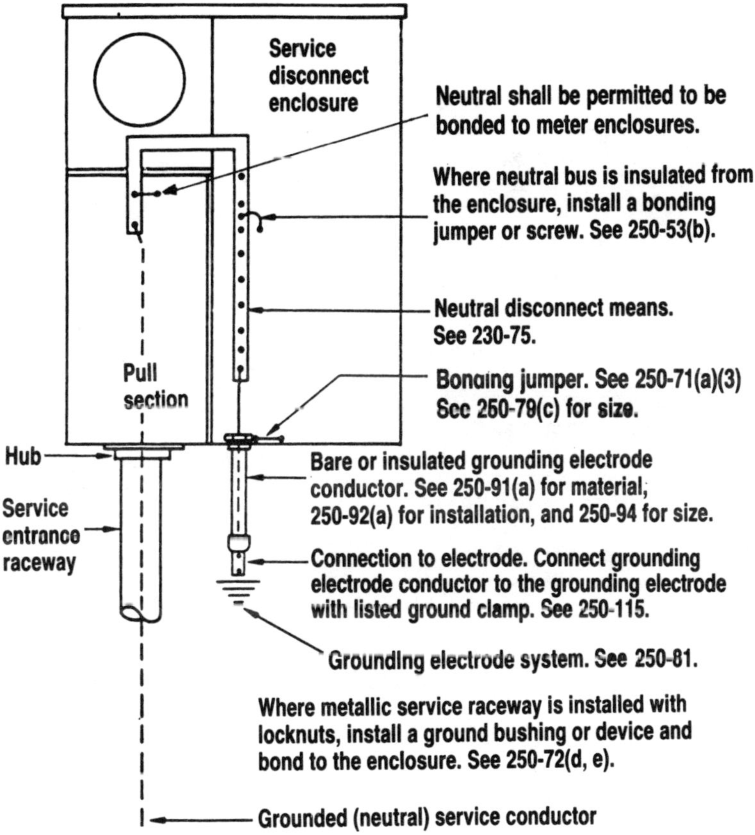

COMBINATION METER AND SERVICE FOR UNDERGROUND MULTIFAMILY DWELLING SERVICE MORE THAN SIX SUBDIVISIONS

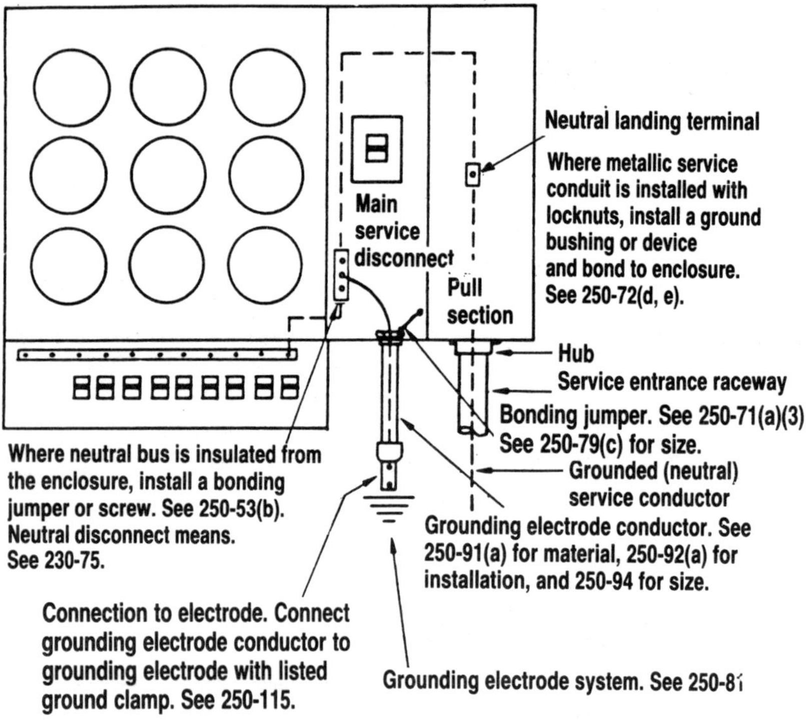

NOTE: *Metallically isolated water piping systems. See 250-80.*

COMBINATION METER AND SERVICE FOR UNDERGROUND MULTI-FAMILY DWELLING SERVICE SIX SUBDIVISIONS OF SERVICE OR LESS

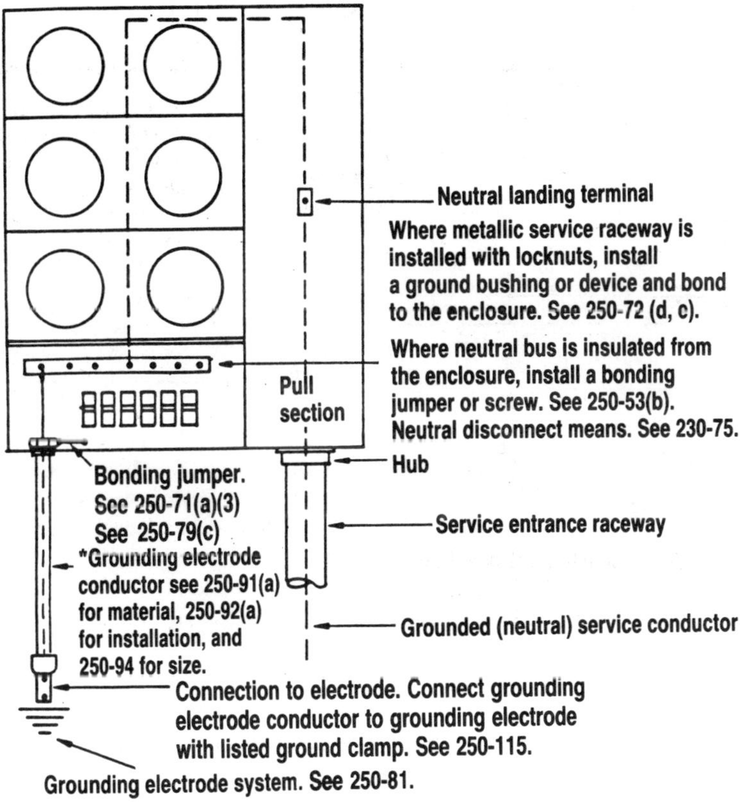

NOTE: *Metallically isolated water piping systems. See 250-80.*

COMBINATION METER AND SERVICE FOR UNDERGROUND COMMERCIAL SERVICE SIX SUBDIVISIONS OR LESS

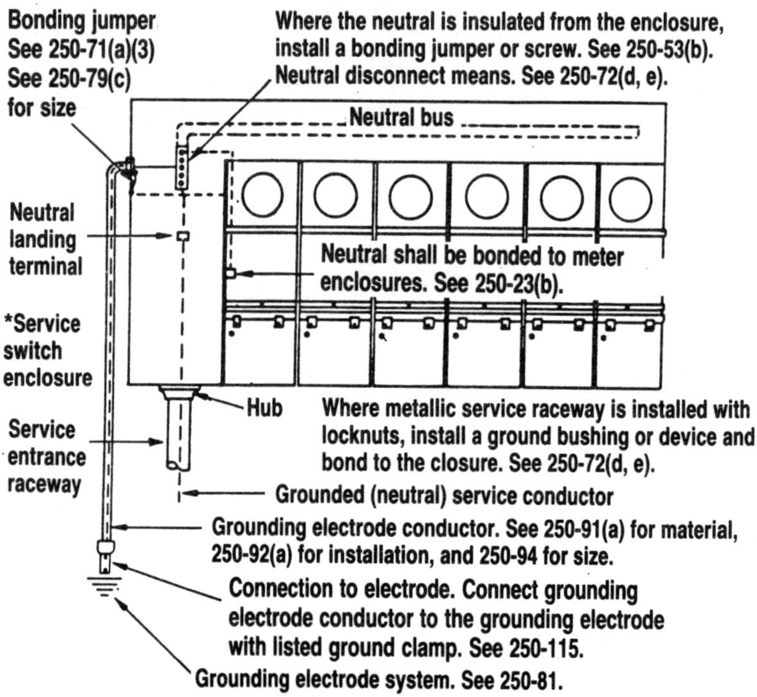

NOTE: *Metallically isolated water piping systems. See 250-80.*

COMBINATION METER AND SERVICE FOR UNDERGROUND COMMERCIAL SERVICE MORE THAN SIX SUBDIVISIONS OF SERVICE

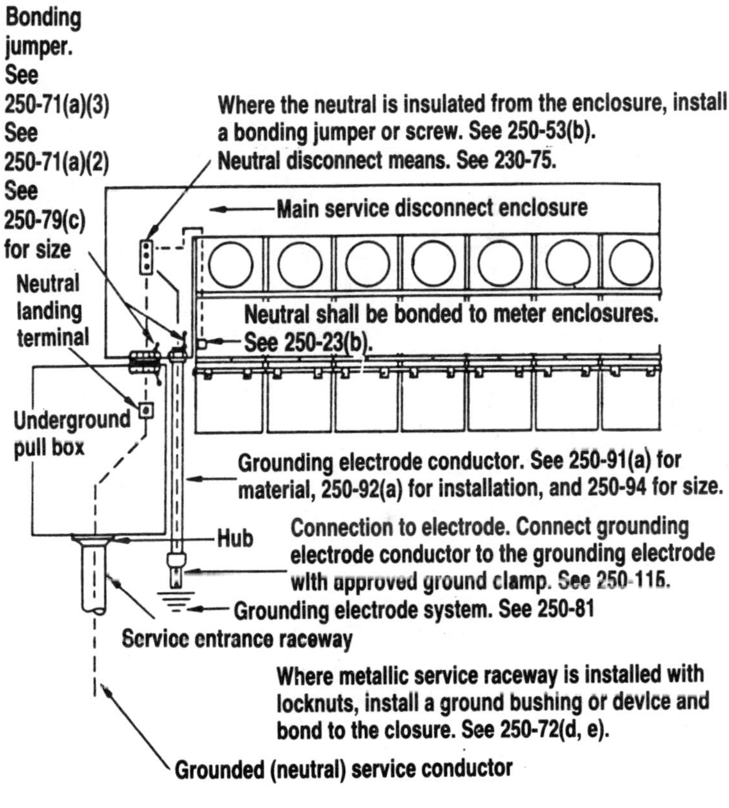

NOTE: *Metallically isolated water piping systems. See 250-80.*

SWITCHBOARD SERVICE SECTION

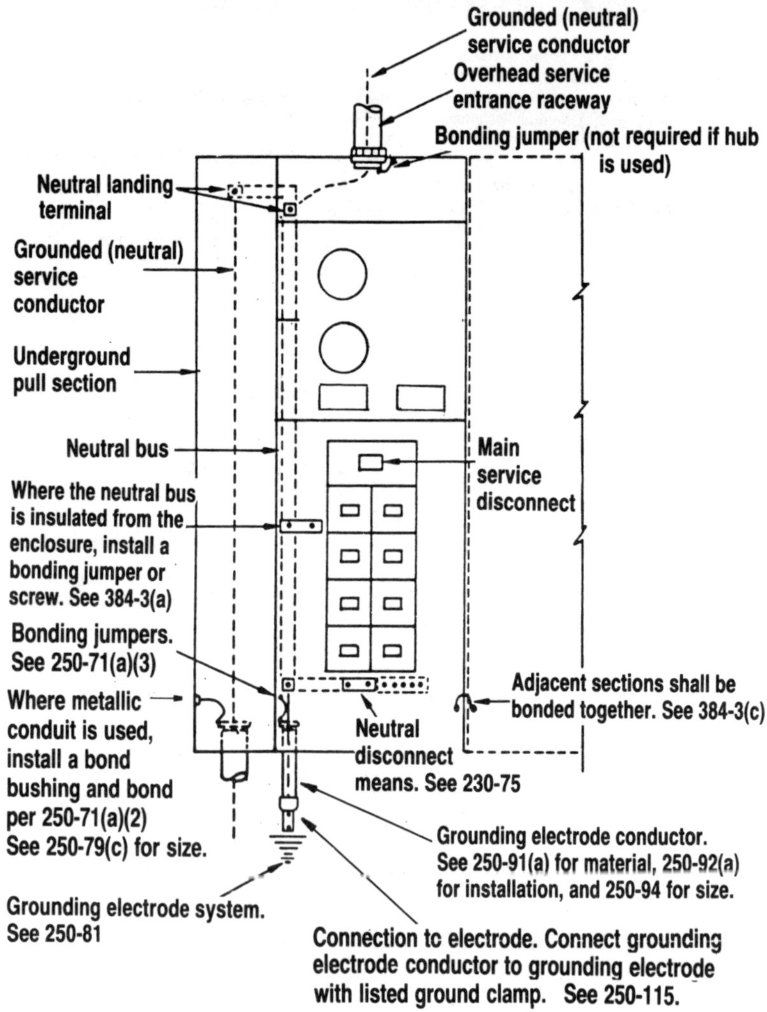

UNDERGROUND SERVICE POST

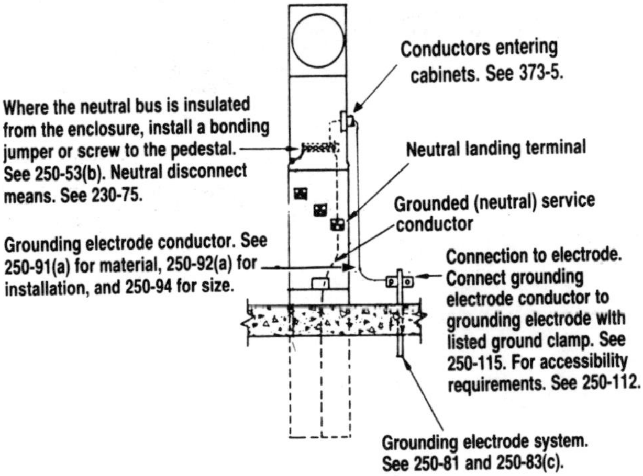

CONCRETE ENCASED ELECTRODE (UFER GROUND)

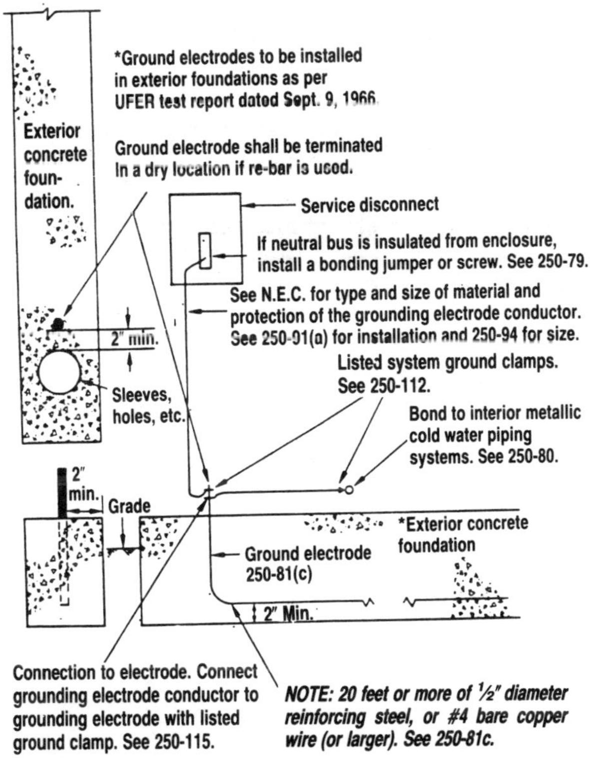

HIGH-IMPEDANCE GROUNDED NEUTRAL SYSTEM, SECTION 250-27

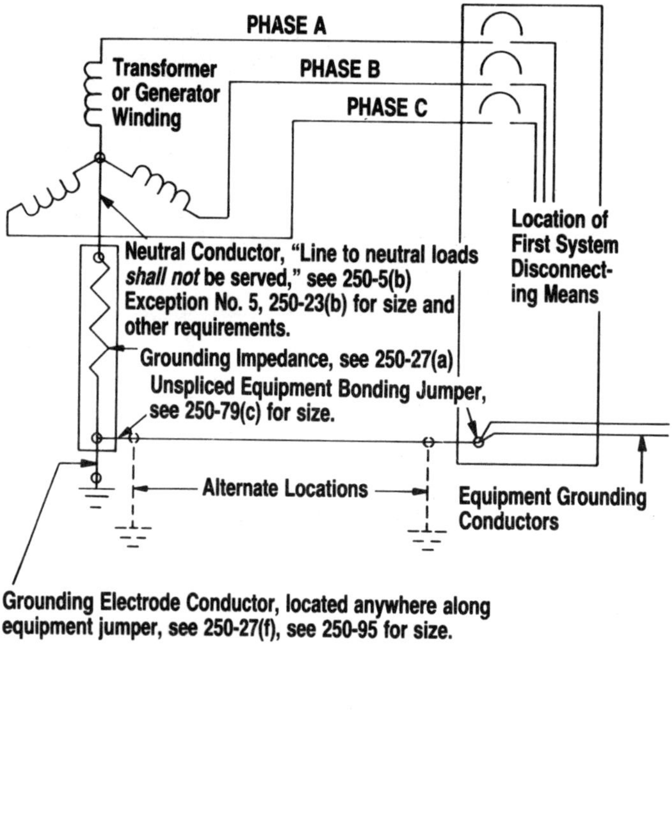

TYPICAL CUSTOMER'S TRANSFORMER BONDING AND GROUNDING

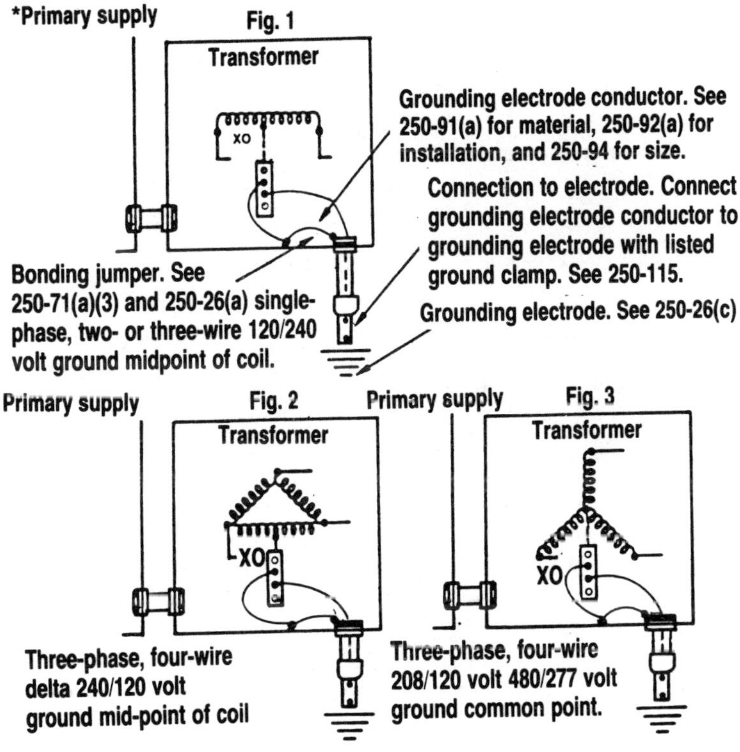

NOTE: Primary windings are not shown for simplicity.

Formulas For Voltage Drop Calculations For Copper Conductors

1. Volts Drop — 1ø

$$\frac{\text{Amperes X Feet X 21.6}}{\text{C.M.}} = \text{Volts Drop} \qquad \%\text{ VD} = \frac{\text{Volts Drop}}{\text{Volts}} \text{ x 100}$$

2. Volts Drop — 3ø

$$\frac{\text{Amperes X Feet X 18.7}}{\text{C.M.}} = \text{Volts Drop}$$

3. Volts drop for 3-wire 120/208 or 277/480 volts circuits derived from 4-wire systems.

   NOTE: This applies only to balanced circuits with double pole switching.

$$\frac{\text{Amperes X Feet X 18.7 X 1.5}}{\text{C.M.}} = \text{Volts Drop}$$

Where:
Amperes = Amperes per terminal of load or demand load if allowed.
Feet = One way distance in feet measured along conductors from the source of supply to the load.
C.M. = Cross sectional area of conductor in circular mils.
Resistance = 10.8 ohms per mil foot for copper & 17.1 ohms per mil foot for aluminum.

CONVERSION TABLE — K.W. TO AMPERES

To find amperes of non-inductive load, multiply KW by Constant.

System	Voltage	Constant
1ø	120	8.33
or	240	4.17
DC	277	3.61
	480	2.08
3ø	120	4.82
	208	2.77
	240	2.41
	480	1.2

LOAD CURRENT CALCULATIONS

Three-Phase Three-Wire Systems. Calculations of loads on three-phase three-wire systems generally involve the addition of single-phase and balanced three-phase loads. Balanced three-wire star or wye loads may be considered as delta loads of equal K.V.A. or line current for the purpose of these calculations. Unbalanced star or wye loads on three-wire systems are very rare. Since all three conductors of a service, feeder or branch circuit are required to be of the same capacity, it is usually necessary to calculate the load only on the conductor which involves the greatest current.

The following procedure simplifies calculations:

(a) Add all balanced three-phase ampere loads directly.

(b) Convert three-phase load into three equal single-phase loads. (The current of each of these equal single-phase loads will be equal to 58% of the three-phase line current.)

(c) Add the single-phase loads on each phase separately. To each of these add the single-phase load derived from the three-phase load in (b) above.

(d) Take the larger two total loads from (c), add them together, and if the unbalance between them is not more than 2 to 1, multiply by 0.87. This will give the total line current. For unbalance of more than 2 to 1 the following multipliers shall be used in place of 0.87.

Not more than 3 to 1—0.90
Not more than 6 to 1—0.94
Not more than 10 to 1—0.96
Over 10 to 1 —1.00

EXAMPLE 1

Equipment:	3ø - Amps. per terminal	Phase A (L_1 to L_2)	Phase B (L_2 to L_3)	Phase C (L_3 to L_1)
Three Phase				
5 HP	15.2			
3 HP	9.6			
10 KW	24.1			
Total 3ø	48.9 x .58 =	28.3	28.3	28.3
Single Phase				
5 KW		20.8		
15 KVA			62.4	
7.5 KVA				31.2
3 KVA		12.5		
		61.6	90.7	59.5
			61.6	
Sum of two largest loads =			152.3	
Multiplied by =			.87	
Line current =			132.5	

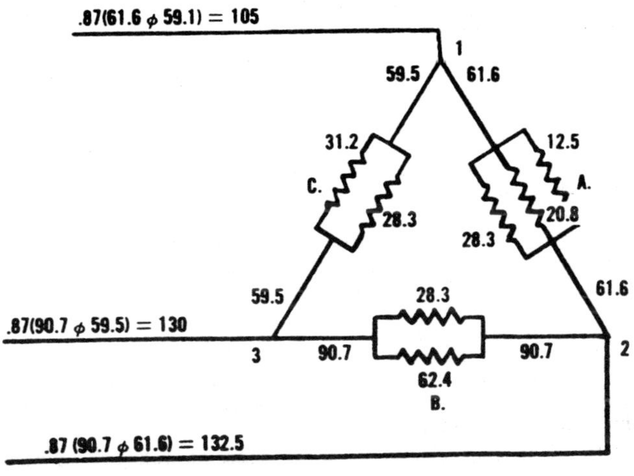

EXAMPLE 2

Equipment:	3ø - Amps. per terminal	Phase A (L_1 to L_2)	Phase B (L_2 to L_3)	Phase C (L_3 to L_1)
Three Phase				
10 HP	28			
5	15.2			
20	54			
15	42			
30	80			
50	130			
Total 3ø	349.2 x .58	202.5	202.5	202.5
Single Phase				
50 KW				208
		202.5	202.5	410.5
				202.2
Sum of two largest loads =				613.
Multiplied by =				.90
Line current =				551.7

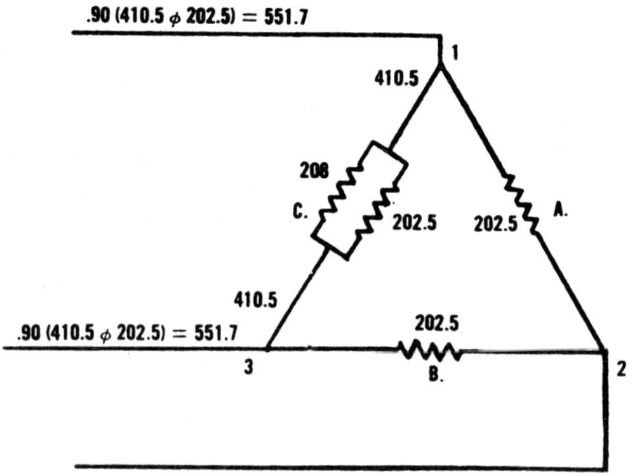

THREE-PHASE FOUR-WIRE SYSTEMS

Loads on four-wire systems usually consist of single-phase loads between line and neutral and balanced three-wire three-phase loads. The line current is equal to the highest single-phase line to neutral load plus the three-phase line load.

If single-phase line-to-line loads are encountered they may be converted to three-phase loads and then added directly to the line current.

Approved Configurations For Plugs And Receptacles

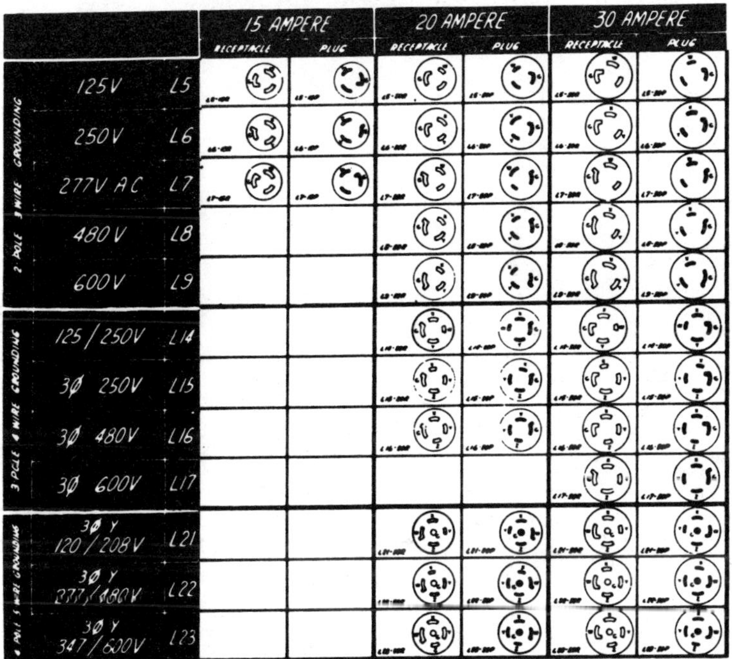

NEMA PUBLICATION NO WD5 FOR DIMENSIONS
68 BY NATIONAL ELECTRICAL MANUFACTURERS ASSOCIATION

Locking Type Plugs and Receptacles for General Purpose 60 Hertz A.C. Use

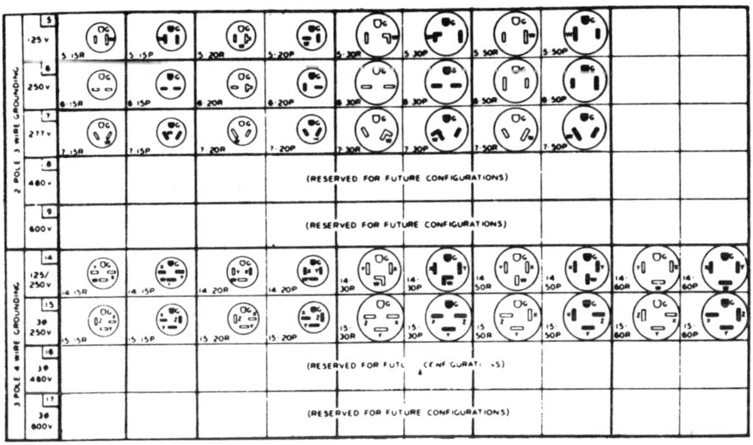

Non Locking Type Plugs and Receptacles for General Purpose 60 Hertz A.C. Use